Innovations in Non-Conventional Energy Sources

This book focuses on exploring and showcasing advancements, breakthroughs, and emerging technologies in the field of energy generation and utilization, particularly those related to non-conventional or alternative energy sources. It delves into various non-conventional energy sources such as solar, wind, geothermal, tidal, and biomass and aims to provide in-depth insights into the technologies associated with these sources. It discusses innovative approaches, recent developments, and technological breakthroughs within the realm of non-conventional energy sources, addressing the environmental benefits of these energies while highlighting their potential to reduce greenhouse gas emissions, combat climate change, and promote sustainable energy practices.

This book:

- Provides real-world examples and case studies of successful non-conventional energy projects.
- Takes an interdisciplinary approach by integrating knowledge from various fields such as engineering, environmental science, economics, and policy, offering a holistic understanding of the subject.
- Emphasizes the environmental benefits of non-conventional energy sources and their contributions to reducing carbon emissions and combating climate change.
- Discusses the policies, regulations, and government incentives that influence the adoption of non-conventional energy sources, explaining how these factors can drive innovation.
- Includes a global perspective by showcasing innovations and projects from different regions and highlighting how various countries are approaching renewable energy.

It serves as a comprehensive resource for researchers, professionals, policymakers, and anyone interested in understanding the latest innovations and trends in the field of non-conventional energy sources, with an emphasis on sustainability, environmental responsibility, and energy security.

Innovations in Non-Conventional Energy Sources

Edited by

Krishan Arora and Himanshu Sharma

CRC Press
Taylor & Francis Group
Boca Raton London New York

CRC Press is an imprint of the
Taylor & Francis Group, an **informa** business

Designed cover image: Shutterstock

First edition published 2026
by CRC Press
2385 NW Executive Center Drive, Suite 320, Boca Raton FL 33431

and by CRC Press
4 Park Square, Milton Park, Abingdon, Oxon, OX14 4RN

CRC Press is an imprint of Taylor & Francis Group, LLC

ISBN: 9781032893945 (hbk)
ISBN: 9781032896748 (pbk)
ISBN: 9781003542568 (ebk)

DOI: 10.1201/9781003542568

Typeset in Sabon
by codeMantra

Contents

Preface

The aim and scope of a book titled *Innovations in Non-Conventional Energy Sources* would likely focus on exploring and showcasing advancements, breakthroughs, and emerging technologies in the field of energy generation and utilization, particularly those related to non-conventional or alternative energy sources.

This book should provide a comprehensive overview of non-conventional energy sources, covering solar, wind, geothermal, tidal, biomass, and other emerging technologies. It should encompass a wide range of innovations in these areas and touch on the basic concepts and terminologies used in these fields. A significant portion of the book would delve into the core algorithms and techniques used in AI and ML, including discussions on supervised learning, unsupervised learning, reinforcement learning, and deep learning. Readers can expect detailed explanations of algorithms such as decision trees, neural networks, and support vector machines.

MATLAB® is a registered trademark of The MathWorks, Inc. For product information,

please contact:
The MathWorks, Inc.
3 Apple Hill Drive
Natick, MA 01760-2098 USA
Tel: 508-647-7000
Fax: 508-647-7001
E-mail: info@mathworks.com
Web: www.mathworks.com

About the editors

Krishan Arora is associated with Lovely Professional University as a Professor with more than 17 years of experience in academics and research. He is currently the Head of the Department of Power Systems in the School of Electronics and Electrical Engineering at Lovely Professional University. He completed his Ph.D. in Electrical Engineering from MM (Deemed to be) University, Ambala. He earned his M.Tech in Electrical Engineering from IKG Punjab Technical University, Punjab, and his B.Tech in Electrical and Electronics Engineering from IKG Punjab Technical University, Punjab. He has published more than 85 research papers in refereed *IEEE*, *Springer*, and *IOP Science* journals and conferences. He has organized several workshops, summer internships, international conferences, special sessions, and expert lectures. He has also published six Indian patents and one copyright. He has supervised more than ten postgraduate theses and more than 30 undergraduate student projects. He has undertaken and completed 15 non-government and consultancy projects. He has attended/participated in 24 national/international online webinars. His areas of expertise include electrical machines, non-conventional energy sources, load frequency control, automatic generation control, and the modernization of smart grids. He has edited five books in different areas of electronics and electrical engineering. He has taught various courses at both UG and PG levels, such as power electronics, non-conventional energy sources, electric drives, induction and synchronous machines, and digital electronics.

Himanshu Sharma earned his Ph.D. in Power Electronics from Punjab Engineering College, Chandigarh. He earned his M.Tech in Electrical Engineering from Punjab Engineering College, Chandigarh. He was associated with Lovely Professional University as an Assistant Professor and has more than 7 years of experience in academics. He is currently an Associate Professor at G. H. Raisoni College of Engineering and Management. He has published more than ten research papers in refereed *IEEE*, *Springer*, and *IOP Science* journals and conferences. He has organized several workshops, summer internships, international conferences, and expert lectures for students. He has supervised more than five postgraduate theses and more than ten undergraduate student projects. He has attended/participated in 20 national/international online webinars. His areas of expertise include power electronics, machine learning, and optimization techniques. He has taught various courses at both UG and PG levels, such as power electronics, non-conventional energy sources, and electric drives. He has edited many books for reputed publishers.

Contributor lists

Vinay Anand
School of Electronics and Electrical Engineering
Lovely Professional University
Punjab, India

Krishan Arora
School of Electronics and Electrical Engineering
Lovely Professional University
Punjab, India

Harpreet Singh Bedi
School of Electronics and Electrical Engineering
Lovely Professional University
Punjab, India

Jumi Bharali
School of Electronics and Electrical Engineering
Lovely Professional University
Punjab, India

Ranjit Bindal
Department of Electrical Engineering
Chandigarh University
Mohali, Punjab, India

Anesu Nicholas Charamba
School of Electrical and Information Engineering
University of the Witwatersrand, 1 Jan Smuts Avenue, Johannesburg, South Africa

Shakti Raj Chopra
School of Electronics and Electrical Engineering
Lovely Professional University
Punjab, India

Amit Kumar Dhir
School of Civil Engineering
Lovely Professional University
Punjab, India

Santosh L. Gaonkar
Department of Chemistry
Manipal Institute of Technology
Manipal Academy of Higher Education
Karnataka, India

Deepika Ghai
School of Electronics and Electrical Engineering
Lovely Professional University
Punjab, India

Mansi Gupta
Department of Computer Science & Engineering
Vidya College of Engineering
Meerut, India

Anil Kumar Jagota
Government Polytechnic College
Hamirpur, Himachal Pradesh, India

Gaurav Kumar
Department of Mechanical Engineering
Vidya College of Engineering
Meerut, India

Hagreaves Kumba
Department of Industrial Engineering,
Durban University of Technology
Durban, South Africa

Denzel Christopher Makepa
Department of Fuels and Energy Engineering
Chinhoyi University of Technology
Chinhoyi, Zimbabwe

Praveen Kumar Malik
School of Electronics and Electrical Engineering
Lovely Professional University
Punjab, India

B. Namratha
Department of Chemistry
The Yenepoya Institute of Arts, Science, Commerce and Management, Yenepoya (Deemed to be University)
Mangaluru, Karnataka

Anesu Nicholas
School of Electrical and Information Engineering
University of the Witwatersrand
Johannesburg, South Africa

Akhil Nigam
Department of Electrical Engineering
Chandigarh University
Mohali, India

Umesh C. Rathore
ABVGIET-Pragatinagar
Shimla, Himachal Pradesh, India

Himanshu Sharma
G. H. Raisoni college of Engineering and Mangement
Nagpur, India

Pushpendra Kumar Sharma
School of Civil Engineering
Lovely Professional University
Punjab, India

Pradeep Singh
ABVGIET – Pragatinagar
Shimla, Himachal Pradesh, India

Tanishk Singhal
School of Electronics and Electrical Engineering
Lovely Professional University
Punjab, India

Suman Lata Tripathi
School of Electronics and Electrical Engineering
Lovely Professional University
Punjab, India

Chapter 1

Enhanced techno-economic optimal sizing of standalone hybrid microgrid power system

Akhil Nigam and Ranjit Bindal

1.1 INTRODUCTION

Today, due to the increase in energy demand, every sector is dependent on energy, with around 85% of energy produced from fossil fuels and renewable energy sources. This situation has led to various challenges such as carbon emissions and rising costs associated with renewable energy, raising concerns in the field of electricity production. In conventional times, electricity depends on a single energy source, such as coal, thermal, gas, or diesel. Consequently, there were challenges in meeting specific energy demands. These resources are considered hazardous due to their environmental impact. Hence, many authors and researchers have shifted their focus to renewable energy. Overall, the energy distribution network is undergoing transformation, driven by the emergence of newly distributed energy sources, including the concept of microgrids.

The term microgrid refers to a small network of distributed energy resources connected to a single power system. DER can include both conventional and non-conventional energy sources. It can perform a properly regulated single-level subsistence for providing islanded or hybrid-tied mode operation. Microgrids have the potential to improve the quality of electricity and overall system efficiency. The concept of microgrid is expected to extend for many years as its technology will not negatively impact resiliency. Different experiments will take place throughout the world to understand the fundamentals of microgrids [1,2]. For various purposes, numerous topologies of microgrids have been studied.

Microgrids are primarily located in the Asia Pacific and North America regions. There is a prediction of an increase in microgrids in the coming years. In 2017, the total production of electricity using microgrids was around 72%. The concept of microgrids is expected to become more significant in remote areas and underserved regions.

The microgrid concept was initially introduced in 1999, aiming to provide a reliable solution for electricity with the interconnection of DER into the grid. By incorporating renewable energy sources like solar, wind, and biogas, the objective was to achieve sustainability improvements. Initially,

DOI: 10.1201/9781003542568-1

it was implemented in an automated manner, apart from the concept of islanding or grid-tied mode [3].

1.2 COMPONENTS OF MICROGRID

A. Generation: A microgrid may comprise both dispatchable and non-dispatchable generation. Dispatchable generation may include conventional energy sources like coal, diesel, and gas, while non-dispatchable generation may include solar, wind, and biogas.
B. Load: A microgrid can contain both types of loads, such as AC loads and DC loads. The nature of the load may be linear or non-linear. Depending on location, loads may be classified into critical and deferrable loads.
C. Energy Management System: The role of the energy management system is crucial in the operation of a microgrid. It ensures the performance of the microgrid through the interconnection of smart meters and communication strategies. It helps synchronize generation and load to meet reliability and economic criteria.
D. Energy Storage System: The energy storage system is primarily used for backing up power supply with the help of different energy resources. It is also used to enhance the stability and efficiency of the microgrid. The EMS helps DER operate without load distortions and ensures frequency regulation while smoothing the output of non-conventional energy sources.
E. Converters: Apart from energy production, there is also a requirement for energy conversion, as there are two modes of microgrids: AC and DC microgrids. Hence, a suitable power converter is required for switching between AC and DC networks.
F. Distributed Energy Resources: This plays a vital role in electricity production, as solar produces DC and wind produces AC. Therefore, according to environmental aspects, DERs have an important role. Since these cannot be fully regulated due to variable weather conditions, they can sometimes become difficult to monitor (Figure 1.1).

1.3 OPTIMAL PLANNING OF MICROGRID

Owing to the continuous growth of distributed generation systems in microgrids, many authors have explored various approaches. The resources may be conventional or non-conventional, but today, an alternative energy source is required for reliable operation. Distributed generation is associated with the utility grid through the interconnection of an intelligent grid. An important aspect to keep in mind is that such configurations must be designed to generate flexible electricity and be capable of transmitting it to

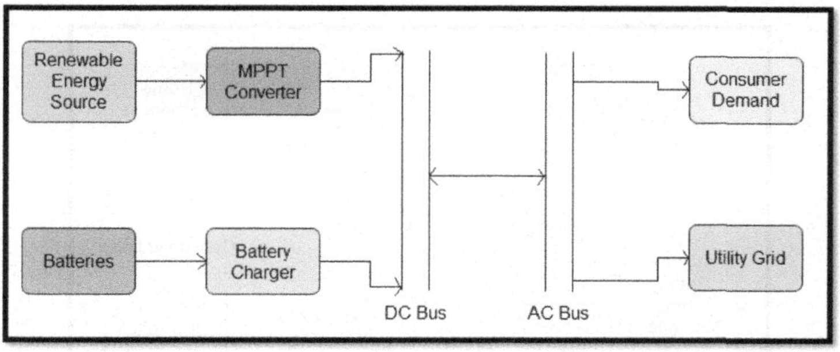

Figure 1.1 Components of microgrid.

remote areas, in particular. The optimal sizing and operation of a microgrid is very important since it uses limited sources of available energy to remain cost-effective [4–6].

For the optimal planning and operation of microgrids, many authors have proposed optimization techniques for both standalone and grid-connected modes [7,8]. Since solar PV cells produce DC, they can be converted into AC for various power applications. Fuel cells are used to store energy in DC, which can be converted by a suitable power converter [9]. For DC microgrid operation, a DC bus must be connected to all its sources and loads, but for AC loads, it cannot always be connected directly. Hence, an AC–DC power converter link is required for energy transmission to the loads [10].

For such operations, there may be some complexity issues because energy transmission requires switching operations. Under these conditions, problems such as voltage sag and voltage swell, along with various distortion issues, may occur, affecting the operation of the microgrid. Many authors have presented issues like voltage stability, feasibility, load balancing, etc. Hence, this indicates complex implementation and reduced reliability [11].

1.4 TYPES OF MICROGRID

A. AC Microgrid: Basically, an AC microgrid consists of distributed generators, an energy storage system, a cluster of loads, and the utility grid. Some sources, such as solar PV cells, small-scale wind turbines, and microturbines, produce DC electricity, which may not be compatible with the utility grid frequency. Sometimes, microturbines produce a higher voltage than the utility grid power frequency. In grid-connected mode, the microgrid utilizes the electric grid while exchanging electricity. Throughout this operation, the frequency must be synchronized with the utility grid. The microgrid acts as a cumulative

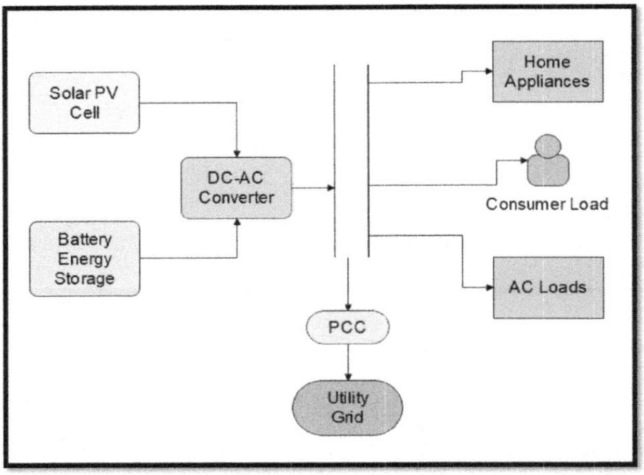

Figure 1.2 AC microgrid structure.

load to the electric grid and is able to regulate harmonics at the point of common coupling (Figure 1.2).

In an AC microgrid, there may be power quality issues due to the adequacy of reactive power compensation, which can lead to voltage sags and voltage swells. Some DGs, such as microturbines and fuel cells, have less inertia and may not be able to provide energy during switching operations. Forecasting techniques are needed to incorporate other intermittent distributed energy sources, like wind turbines, to ensure they can operate under such conditions. In some applications, an AC microgrid can be integrated with high-frequency links, making such configurations suitable for aerospace applications [12].

B. DC Microgrid: In recent years, there have been changes in the field of semiconductor devices, which are now being integrated with power electronic devices and renewable energy sources like solar, wind, and geothermal. Various applications, such as mobile devices, laptops, mobile phones, and televisions, utilize a DC power source from DC batteries or other external DC power sources. Based on these applications, a DC microgrid has been introduced. Many authors have presented their work on DC microgrids and highlighted numerous benefits. These configurations reduce issues such as energy conversion loss, reactive power compensation, and synchronization with the electric grid. The DC microgrid can operate in both grid-connected and islanded modes (Figure 1.3).

C. Hybrid Microgrid: The authors have introduced the concept of a hybrid microgrid, which may combine AC and DC microgrids, based on their study of previous research. This approach can provide a smart solution

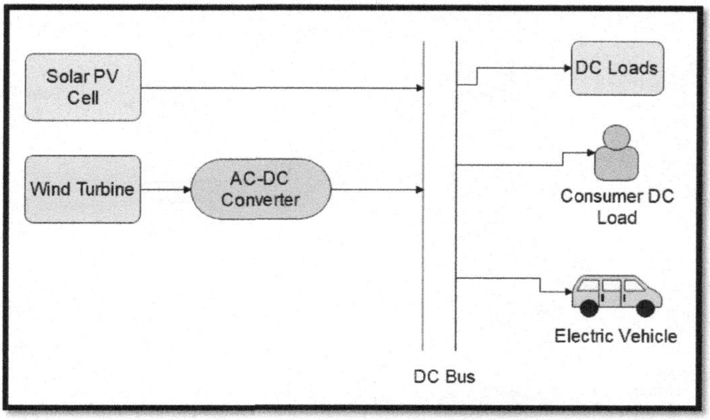

Figure 1.3 DC microgrid structure.

for distributed transmission networks. Sometimes, problems arise due to the unavailability of resources; however, with the help of a battery energy storage system, electricity can be utilized. The authors have explored different hybrid configurations and conducted comparisons [13].

The advantage is that the two networks, AC and DC, can be combined in the same utility grid, facilitating the direct integration of both AC and DC-based DG systems. Such configurations can be associated with emerging alternative energy sources or electric vehicles, which represent advanced techniques. Hence, this hybrid microgrid configuration offers an optimal solution for integrating advanced grids with traditional distribution networks (Figure 1.4).

There are various advantages to employing hybrid microgrid infrastructure, which are beneficial from the perspective of smart grids in distributed energy networks:

A. Amalgamation: By combining AC and DC microgrids, they can directly interface with basic elements. This interfacing reduces energy losses at different energy conversion stages.
B. Synchronization of Hybrid Microgrid: Proper synchronization with the load allows it to produce electricity according to consumer demand. In this process, an effective controlling approach is essential.
C. Stability: Hybrid microgrids can provide stable responses; however, switching operations can introduce fluctuations that affect their performance.
D. Feasibility: This hybrid combination can offer an economically feasible response, although it may sometimes increase maintenance costs. Efficiency can be reduced, so selected RES must be interfaced to ensure reliable operation.

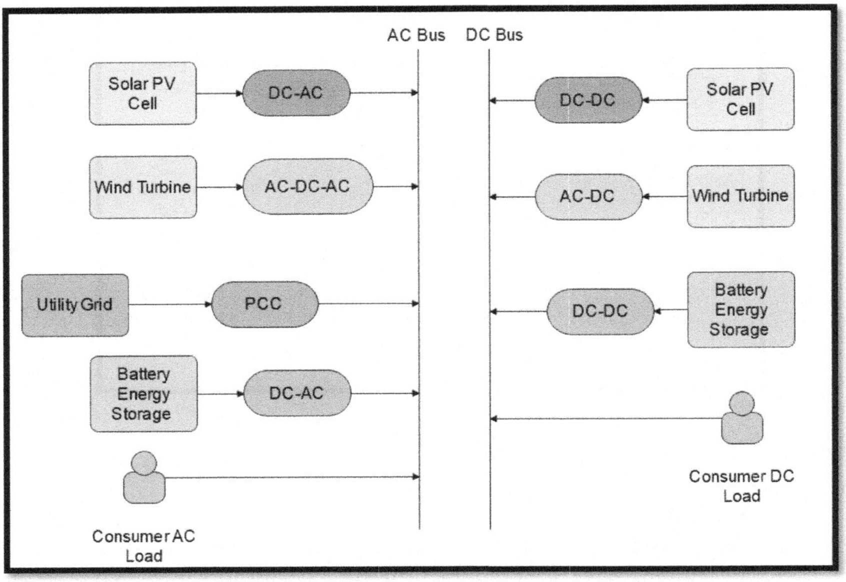

Figure 1.4 Hybrid microgrid.

Similarly, hybrid microgrids may introduce some disadvantages that affect performance, such as reliability, complexity, and safety issues. Employing multiple components makes hybrid microgrids complex, potentially increasing costs. Additionally, energy transmission through power converters can introduce distortions that reduce efficiency and reliability at various energy stages. Therefore, various controlling approaches have been introduced for this operation.

1.5 OPTIMIZATION APPROACHES FOR OPTIMAL SIZING OF HYBRID MICROGRID

Many authors have proposed various optimization techniques for enhancing the proper location of DG sources and attaining the optimal sizing of microgrid components. It is necessary to locate the components in a microgrid whether it is operating on AC supply or DC supply. Authors have proposed an evolutionary-based optimization method for a hybrid microgrid containing solar, wind, and diesel energy sources for the minimization of carbon gas emissions and costs [14]. One author proposed a genetic algorithm for a solar–wind hybrid system to reduce energy losses [15]. Another PSO optimization technique has been discussed for the optimal sizing of an alternative energy-based microgrid [16].

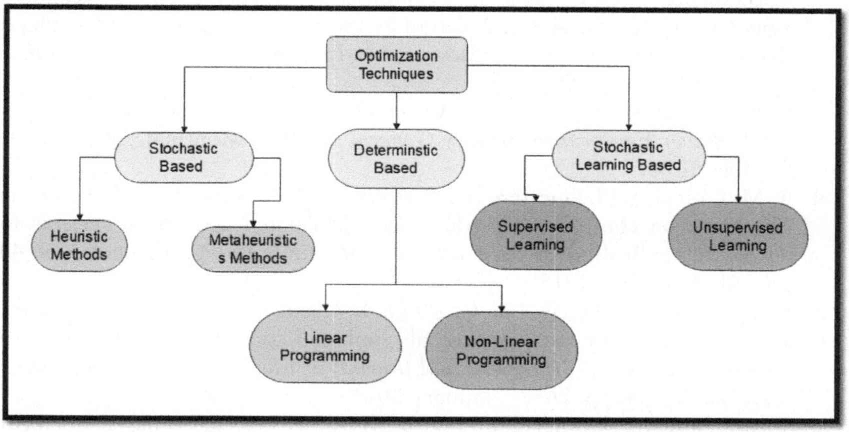

Figure 1.5 Classification of optimization techniques.

A simulated annealing-based technique was proposed by an author for a hybrid system to achieve optimal sizing and cost reduction [17]. Furthermore, a multi-objective PSO was applied for determining the optimal sizing of hybrid microgrid operations [18]. To achieve higher responsiveness, authors began combining hybrid optimization techniques, which may be heuristic or metaheuristic in nature. This is the basic classification of optimization techniques. For different operations, a specific optimization technique is required.

Figure 1.5 shows the classification of optimization techniques, categorized according to specific objective functions. A linear programming-based optimization technique is generally better than a non-linear programming-based technique. Evolutionary-based techniques are basically types of metaheuristic techniques [19]. Based on different optimization approaches, various objectives have been identified and analyzed for the operation of hybrid microgrids. Different aspects like the cost of energy, life cycle cost, loss of load, and power supply probability have been studied by various researchers. Furthermore, authors have utilized different software tools for optimization approaches to determine the best optimal solution for hybrid microgrids, along with their advantages and disadvantages [20].

REFERENCES

1. M. H. Saeed, W. Fangzong, B. A. Kalwar, and S. Iqbal, "A review on Microgrid's challenges and perspectives", *IEEE Access*, vol. 9, pp. 166502–166517, December 2021.

2. M. Sandelic, S. Peyghami, A. Sangwongwanich, and F. Blaabjerg, "Reliability aspect in microgrid design and planning: status and power electronics-induced challenges", *Renewable and Sustainable Energy Review*, vol. 159, pp. 5, pp. 112–127, May 2022.
3. P. Piagi, and R. H. Lasseter, "Autonomous control of microgrids". *2006 IEEE Power Engineering Society General Meeting*, Montreal, QC, Canada, October 2006.
4. A. M. Bouzid, J. M. Guerrero, M. Bouhamida, P. Sicard, and M. Benghanem, "A survey on control of electric power distributed generation systems for microgrid application", *Renewable ad Sustainable Energy Reviews*, vol. 44, pp. 751–766, April 2015.
5. W. Gu, Z. Wu, R. Bo, W. Liu, G. Zhou, W. Chen, et al., "Modeling, planning and optimal energy management of combined cooling, heating and power microgrid: a review", *International Journal of Electrical Power and Energy Systems*, vol. 54, pp. 26–37, January 2014.
6. S. Mohammadi, S. Soleymani, and B. Mozafari, "Scenario based stochastic operation management of microgrid including wind, photovoltaic, micro-turbine, fuel cell and energy storage devices", *International Journal of Electrical Power and Energy Systems*, vol. 54, pp. 525–535, January 2014.
7. Q. Jiang, M. Xue, and G. Geng, "Energy management of microgrid in grid-connected and standalone modes," *IEEE Transactions on Power Systems*, vol. 28, no. 3, pp. 3380–3389, August 2013.
8. M. Mohammad, "Optimal operation management of a typical microgrid as grid connected in power systems using fuzzy sliding-mode control (FSMC) approach," *World Applied Sciences Journal*, vol. 28, no. 4, pp. 440–448, 2013.
9. G. T. Heydt, "The next generation of power distribution systems," *IEEE Transactions on Smart Grid*, vol. 1, no. 3, pp. 225–235, December 2010.
10. S. Khosrogorji, M. Ahmadian, H. Torkaman, and S. Soori, "Multi-input DC/DC converters in connection with distributed generation units – a review," *Renewable and Sustainable Energy Reviews*, vol. 66, pp. 360–379, December 2016.
11. A. Banerji, D. Sen, A. K Bera, D. Ray, D. Paul, B. Anurag, and S. K. Biswas, "Microgrid: a review", *IEEE Global Humanitarian Technology Conference: South Asia Satellite*, Trivandrum, India, pp. 27–35, October 2013.
12. K. Arora, G. P. Joshi, M. Ragab, M. Rawa, A. H. Milyani, R. F. Mansour, and E. Yang, "Bilateral contract for load frequency and renewable energy sources using advanced controller", *Computers, Materials & Continua, Tech Science Press*, vol. 73, no. 2, pp. 3165–3180, 2022. https://doi.org/10.32604/cmc.2022.026966
13. E. Planas, J. Andreu, J. I. Garate, I. M. de Algeria, and E. Ibarra, "AC and DC technology in microgrids: a review", *Renewable and Sustainable Energy Reviews*, vol. 43, pp. 726–749, 2015.
14. J. L. Bernal-Agustín, R. Dufo-López, and D. M. Rivas-Ascaso, "Design of isolated hybrid systems minimizing costs and pollutant emissions," *Renewable Energy*, vol. 31, no. 14, pp. 2227–2244, 2006.
15. S. M. M. Tafreshi, H. A. Zamani, S. M. Ezzati, M. Baghdadi, and H. Vahedi, "Optimal unit sizing of distributed energy resources in microgrid using genetic algorithm," in *Proceeding of ICEE*, Isfahan, Iran, pp. 836–841, May 2010.

16. M. Pirhaghshenasvali, and B. Asaei, "Optimal modeling and sizing of a practical hybrid wind/PV/diesel generation system," in *Proceedings of the 5th Power Electronics, Drive Systems and Technologies Conference (PEDSTC)*, Tehran, Iran, pp. 506–511, February 2014.
17. R. Mahajan, H. Sharma, K. Arora, G. P. Joshi, and W. Cho, "Comparative analysis of the gazelle optimizer and its variants", *Heliyon Journal*, https://doi.org/10.1016/j.heliyon.2024.e36425
18. H. R. Baghaee, M. Mirsalim, and G. B. Gharehpetian, "Multi-objective optimal power management and sizing of a reliable wind/PV microgrid with hydrogen energy storage using MOPSO," *Journal of Intelligent and Fuzzy Systems*, vol. 32, no. 3, pp. 1753–1773, January 2017.
19. R. Mallipeddi, and P. N. Suganthan, "Differential evolution algorithm with ensemble of populations for global numerical optimization," *OPSEARCH*, vol. 46, no. 2, pp. 184–213, 2009.
20. Sk. A. Shezan, I. Kamwa, Md. F.n Ishraque, S. M. Muyeen, K. N. Hasan, R. Saidur, S. M. Rizvi, Md. Shafiullahn and F.A. Al-Sulaiman, "Evaluation of different optimization techniques and control strategies of hybrid microgrid: a review", *Energies*, vol. 16, February 2023.

Chapter 2

Study of battery management system (BMS) on fire mitigation techniques of electric vehicles

Ranjit Bindal and Akhil Nigam

2.1 INTRODUCTION

Globally, the disruption in the mobility sector is leading to a faster adoption of a shared, connected, autonomous e-mobility future. The biggest hurdle to the rapid adoption of electric vehicles (EVs) is the reliability and performance metrics of EVs at a competitive price point compared to internal combustion engine-based vehicles. The fierce competition among original equipment manufacturers (OEMs), the lack of Indian standards prior to 2022, and insufficient quality control in battery manufacturing have heightened the probability of fire incidents. The recent surge in two-wheeler EV battery fires has not only hindered the faster adoption of EVs in India but has also instilled fear among the public due to these sporadic incidents leading to loss of life and property. The challenge of achieving good performance in EV two-wheelers while maintaining adequate quality at a reasonable cost is a significant issue, alongside the knowledge gap on the part of OEMs. A battery serves as an efficient means of storing a significant amount of electrical energy over an extended period. To ensure the safe and consistent operation of the battery pack, a battery management system (BMS) is employed as a control unit. The main purpose of the BMS is to protect and preserve the battery. This involves crucial tasks such as monitoring individual cells for safety, implementing cell balancing techniques to address issues like aging, and supervising the overall battery pack. The BMS is responsible for taking corrective actions in response to any abnormal conditions within the system. Additionally, the BMS regulates the system temperature, as it impacts the power consumption profile, ensuring appropriate measures are in place to control the temperature effectively. Lithium-ion batteries have been the driving force behind the rapid adoption of EVs due to their high energy density, cycle life, and minimal memory effect [1]. However, they have poor heat dissipation capability and are susceptible to thermal runaway. Currently, there is a lack of dedicated BMS standards for various applications such as large-scale energy storage, small

DOI: 10.1201/9781003542568-2

Table 2.1 Report on two-wheeler EV fire incidents in India

Date	*Brand*	*Place*	*Loss*	*Remarks*
Feb 23	Ather 450x	Bengaluru	Battery safe, scooter burnt	Inactive
Dec22	Unknown	Mumbai	Scooter burnt	Inactive
Dec 22	Pure EV	Mumbai	Scooter burnt	Inactive
Oct 22	Unknown	Noida	Scooter burnt	Driving
Oct 22	Batt:RE	Mumbai	Battery explodes, boy killed	Charging
Oct 22	Batt:RE	Mumbai	16 EVs burnt, several injured	Inactive
Sep 22	Unknown	Hyderabad	8 persons dead	Charging
Aug 22	Unknown	Bhiwandi	Scooter burnt	Charging
Jun 22	Oakinawa	Mangalore	34 EVs burnt	Unknown
May 22	Ola	Pune	EV burnt	Inactive
Apr 22	Jitendra EV	Nashik	20 EVs burnt	Inactive

devices, and electric vehicles. Recognizing the significance of such systems and their critical role in ensuring the safe functioning of EVs, the purpose of this paper is to conduct a study specifically focused on BMS in transportation applications, with an emphasis on standardization efforts. The paper undertakes a thorough assessment of BMS components, system architectures, and safety risks associated with BMS operations. Furthermore, it provides commentary on relevant technical standards related to BMS to facilitate the development of new standards in this field. Table 2.1 clearly shows the variability in fire incidents involving EV two-wheelers concerning operating conditions and indicates that root cause analysis by OEMs is either lacking or the information is not accurately shared in the public domain. Except for one case involving Ather, where the BMS reportedly prevented any damage to the battery with the help of other control strategies, in the rest of the cases, all safety mechanisms failed.

2.2 ANALYSIS OF EV FIRE

The major reasons for two-wheeler EV fire incidents are attributable to poor battery pack design, inadequate quality control of imported lithium cells, insufficient redundancies and safety mechanisms, the absence of fire suppression systems, lack of telemetry, inadequate BMS, and the use of unauthorized or inadequate charging infrastructure. The battery design must account for not only mechanical strength in the event of a collision but also ensure adequate spacing between cells to prevent thermal propagation and thermal runaway. Most OEMs manufacture battery packs using imported lithium-ion cells, which need to be carefully examined to eliminate defective cells [2] and ensure compliance with existing standards. Cell balancing is an important aspect of battery pack assembly and testing. Unlike four-wheeler EVs, which have fire detection and suppression systems to allow sufficient time for passenger evacuation in the event of a battery fire, most two-wheeler EVs lack an active fire suppression system. Most OEMs depend on the conduction and convection of heat from the battery pack, which is grossly inadequate in the event of thermal runaway and can lead to battery pack explosions. However, a few OEMs have begun manufacturing battery packs that incorporate passive fire suppression systems based on phase change materials and venting mechanisms. The BMS is the heart of the lithium battery, and a poorly designed or entirely absent BMS, which is essential for charging and discharging, can lead to battery pack explosions. A safe charging system that interacts positively with the BMS is key to the safe charging of EV batteries. This includes ensuring proper grounding in home electrical supplies and using only specified chargers. Telemetry is mostly absent in two-wheeler EVs, as it is costly and requires significant cloud computing and storage, which are vital for forensic analysis and real-time alerts to the EV user. Prior to 2020, regulations for manufacturing two-wheeler EVs were lax, and their effectiveness in eliminating subpar cells for battery pack fabrication was poor, leading to non-standardization and a flood of low-quality two-wheeler EVs, resulting in an increase in fire incidents. AIS 156 [3] clearly established a standard that can be used by OEMs as a benchmark for producing safe and reliable products with a focus on standardization. Some additional factors and considerations related to the causal relationship with fire incidents are as follows:

- Faulty Electrical and Mechanical Design: Poorly designed electrical and mechanical components can lead to issues such as inefficient energy transfer, inadequate cooling, or structural weaknesses that impact the overall performance and safety of the BMS.
- Production Quality Deficiencies with the Cells: Battery cells may suffer from production defects, such as inconsistent cell chemistry, improper sealing, or electrode misalignment. These issues can affect the demonstration, capacity, and lifespan of the battery and should be thoroughly tested during production.
- Flawed or No Battery Monitoring: Accurate monitoring of battery specifications such as voltage (V), current (A), temperature, and state of charge (SoC) is crucial for effective battery management. A flawed or non-existent monitoring system can lead to inaccurate readings, inadequate protection mechanisms, and potentially hazardous situations.

- Defective Charging/Discharging Process: Improper charging and discharging protocols can lead to overcharging, undercharging, or excessive temperature rise, which can degrade battery performance, reduce lifespan, or even cause safety hazards. Robust charging and discharging algorithms should be implemented and tested.
- Lack of Testing for Battery and BMS: Comprehensive testing should be performed on both the battery cells and the BMS to ensure their performance, safety, and reliability. This includes validating cell characteristics, BMS functionality, communication protocols, and protection mechanisms.
- Inadequate Cell/PackLevel Testing: Testing individual cells and evaluating their performance under different operating conditions is important for identifying weak or faulty cells. Additionally, pack-level testing is necessary to verify the overall performance of the battery pack, including its energy capacity and thermal behavior.
- Cycle Life (Charge/Discharge Cycling): Batteries undergo charge and discharge cycles during their lifetime. Testing the battery's ability to handle a specified number of cycles is crucial for understanding its degradation rate, capacity retention, and performance over time.
- Cycle Life (Simulated Drive/Usage Profiles): Simulating real-world usage profiles and applying them to the battery pack can help assess the battery's performance under typical operating conditions, including different discharge rates, temperature variations, and cycling patterns.
- Hybrid Pulse Power Characterization: Hybrid pulse power characterization testing involves subjecting the battery to dynamic load profiles that simulate high-power applications. This helps evaluate the battery's response to sudden power demands, assess its voltage sag behavior, and ensure its suitability for high-power applications.
- Overcharge: Testing the battery's response to overcharge conditions is crucial for verifying the effectiveness of protection mechanisms and ensuring that the battery can safely handle excessive charging voltage.
- Short Circuit: Evaluating the battery's response to a short circuit is important for assessing its safety features, including the ability to detect and protect against short circuit events without causing thermal runaway or other hazardous conditions.
- Mechanical Shock: Batteries can experience mechanical shocks during transportation or usage. Testing the battery's resilience to mechanical shocks helps ensure its structural integrity, reliability, and safety.
- Thermal/Humidity Cycling: Thermal cycling tests involve subjecting the battery to temperature variations, while humidity cycling tests assess its performance under humid conditions. These tests help evaluate the battery's stability, aging characteristics, and the impact of environmental conditions on its performance and safety.

- Thermal Shock: Thermal shock testing subjects the battery to rapid temperature changes to evaluate its ability to withstand sudden temperature variations without compromising its functionality or safety.
- Altitude: Testing the battery's performance and safety at different altitudes is important, especially for applications in high-altitude environments. It helps ensure that the battery can operate reliably and safely under reduced atmospheric pressure conditions.

2.3 CONCEPT OF THERMAL RUNAWAY

The maximum incidents of EV fires are attributable to thermal runaway of lithium battery packs [5]. The energy density of a typical Li-ion two-wheeler battery unit weighing about 7 kg is approximately 150 Wh/kg, which translates into approximately 1 kWh of total energy. This can be estimated to be equivalent to the energy produced by more than 40 L of petrol, explaining the violence and magnitude of these fire incidents in two-wheelers. Thermal runaway is an uncontrolled reaction characterized by a temperature increase of more than 10°C per minute, heavy smoke, and powerful jet flames [4]. Various passive design measures and controls [5] are employed while assembling a battery pack, such as cell spacing and BMS control over all aspects of charging and discharging, along with thermal sensors. However, active measures are almost non-existent in the two-wheeler segment. In the context of thermal runaway in lithium-ion batteries, the conventional fire triangle, which typically consists of fuel, oxygen, and an ignition source, may not fully apply. Thermal runaway in Li-ion batteries is characterized by an internal chemical reaction that releases heat and can sustain itself without requiring external oxygen or a visible flame. Therefore, the management approach for thermal runaway in Li-ion batteries differs from that of a conventional fire. In thermal runaway events, simply starving the reaction of oxygen may not be effective until the exothermic reaction is complete. Instead, the primary focus is on removing the heat generated by the reaction or dispersing the fuel to mitigate the thermal runaway event. This can be achieved through various means, such as cooling the battery with water or other effective cooling methods. Heat removal is crucial because it helps prevent further escalation of the thermal runaway event and mitigates the risk of thermal propagation to neighboring cells. Effective cooling mechanisms can dissipate heat and prevent the battery from reaching critical temperatures. In some cases, it may also be necessary to disperse the fuel source to halt the thermal runaway event. This can involve methods such as physically separating or isolating the battery from other cells or preventing the spread of the reaction through containment measures. There are three stages of thermal runaway [6] (Figure 2.1).

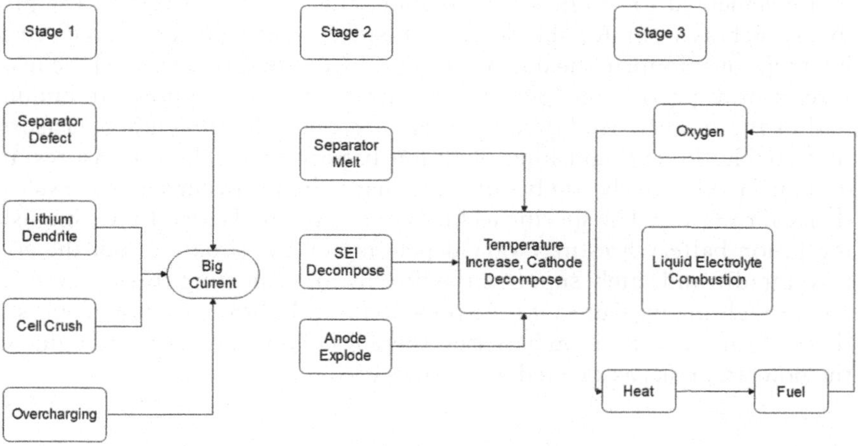

Figure 2.1 Three-stage operation of thermal runaway response.

The first stage is the increase in temperature of the lithium-ion battery, which may be due to overcharging, internal/external short circuits, mechanical damage, formation of dendrites from charging at very low temperatures, failure of the separator, etc. The second stage is highly exothermic, leading to temperature rises from 90°C to 180°C and decomposition of the cathode with a large release of oxygen. During stage 3, combustion of the organic electrolyte occurs, resulting in high volatility, a rapid increase in temperature, jet of gases, and explosions. If it becomes unfeasible to prevent the temperature of a Li-ion battery from rising, it will inevitably lead to a thermal runaway phenomenon, resulting in a fire within the cell. The occurrence of such a thermal runaway response depends on the battery's SoC, with higher charge levels leading to a lower temperature threshold for the reaction to begin. The composition of Li-ion battery materials also impacts the temperature at which thermal runaway circumstances initiate, with LCO batteries having the lowest temperature threshold. Other factors that affect the onset of thermal runaway include battery enclosure, state of health (SoH) of the battery, capacity, and ambient temperature. The combustion characteristics of the battery are crucial in assessing the danger associated with a cell fire, providing fundamental data for evaluating the associated risks. During battery fire tests, LCO and LFP cells exhibited the highest power, while NMC cells demonstrated the lowest power. Additionally, as observed in reference, the timing of the thermal runaway occurrence influenced the amount of energy produced during a fire, specifically the maximum heat release rate. A higher battery SoC resulted in a shorter time to reach a higher value of maximum heat release rate. To mitigate the effects of thermal runaway, various preventive measures can be implemented to minimize the flammability of the battery. Safety protocols such as separator shutdown and battery venting can be employed to interrupt or minimize the damage caused by thermal runaway. These measures aim to prevent the spread of the fire and excessive pressure buildup within the battery. Additionally, there are post-TR firefighting measures that enable the extinguishing of Li-ion battery fires. These measures are designed to effectively combat fires that have already occurred as a result of thermal runaway. The specific techniques and methods used for extinguishing Li-ion battery fires may vary depending on the situation, but the goal is to rapidly and safely suppress the fire. By implementing these preventive and firefighting measures, it is possible to limit the adverse consequences of thermal runaway in Li-ion batteries, thereby enhancing safety and reducing the potential risks associated with battery fires.

2.4 BMS-BASED MITIGATION TECHNIQUES

The BMS plays a crucial role in ensuring safe operation, optimizing energy usage, controlling charging functionality, and overall monitoring of an EV. It is a key component that ensures the efficient and reliable performance of the vehicle's powertrain system [7]. The BMS is the heart and soul of EV operation, dictating the safety, performance, and reliability of the vehicle. In addition, it is responsible for evaluating SoC and SoH, taking into account the effects of temperature and aging/cyclic effects on the cells' electrochemistry. The BMS manages all functions related to charging and discharging of the battery pack, such as overcharging protection, temperature monitoring of cells, undervoltage protection, short circuit protection, cell balancing, isolation of high voltage from the EV, and implementing isolated communication within the vehicle network. These functions are vital to prevent overheating of cells, which can lead to thermal runaway [8]. Figure 2.2 provides a visual representation of the powertrain system, which includes the BMS as an integral part of the overall EV operation.

The battery-powered EV system structure comprises the traction battery as the sole power source, characterized by its significant capacity and high-power output. The battery operates in two main modes: charging and discharging. In discharging mode, the battery supplies power to the electric motor, which transforms electrical energy into mechanical energy. The mechanical drive system then conveys this rotational energy to the vehicle's wheels, propelling the vehicle forward.

Additionally, the battery fulfills the power demands of the vehicle, including air conditioning, sensing systems, communication devices, infotainment systems, and other on-board requirements. BMS safety is a paramount concern in the battery industry. Numerous codes and standards have been developed and implemented for various applications to ensure battery safety. Despite batteries being a convenient energy storage

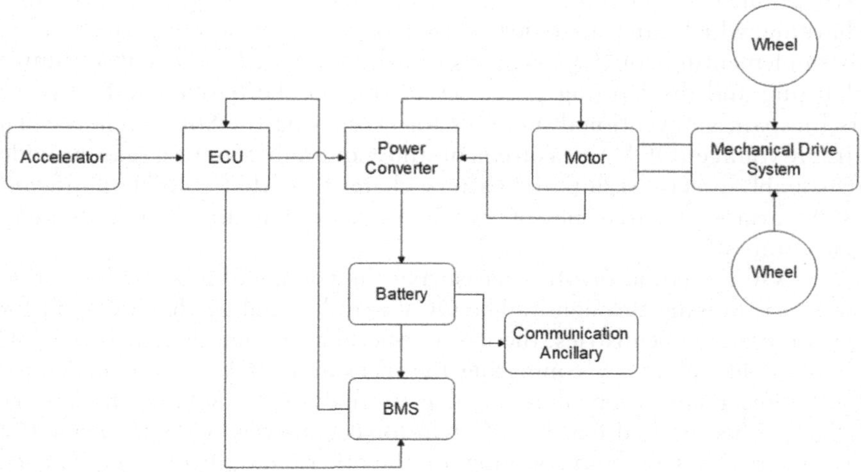

Figure 2.2 Typical powertrain layout in EV.

solution, accidents still occur within battery infrastructure. Consequently, it is crucial to thoroughly consider all design scenarios to mitigate risks and enhance safety measures associated with battery systems. To ensure the battery system operates within the limits set by the BMS, the BMS must have control over the battery systems through bus communication. The BMS is responsible for maintaining compliance with safety procedures by regulating the on/off requirements of main contactors and monitoring voltage, current, and temperature profiles. Unless explicitly stated in the manufacturing procedure, the BMS supports battery systems by providing technical, operational, and safety control. If the BMS receives energy from an external power source, it must accurately record and consider this energy input in its safety functions to ensure effective monitoring and control of the battery system.

2.5 ALGORITHM FOR SoC ESTIMATION

The SoC of a battery refers to a relative measure of the energy stored within it. It is determined by calculating the ratio between the amount of charge that can be extracted from the battery at a particular moment and its total capacity. Accurately estimating the SoC is crucial because BMSs rely on this information for several purposes. Firstly, the SoC estimate informs the user about the expected usage of the battery until the next recharge [7]. This helps users plan their activities and ensures they understand how much energy is available. Furthermore, maintaining the battery within a safe operating window depends on accurate SoC estimation. BMSs monitor the SoC to prevent overcharging or deep discharging, which can lead to degradation or even destruction of the battery. By implementing control strategies based on SoC, the BMS can optimize charging and discharging processes to protect the battery and enhance its lifespan. Conventional methods for estimating the SoC, such as open circuit voltage (OCV) measurement and Coulomb counting, can provide reasonably precise results for battery chemistries that exhibit significant OCV variations across the SoC range, provided that current measurement is accurate [9].

The OCV method involves measuring the voltage of a battery when no current is flowing through it. The OCV is influenced by the SoC, and for certain battery chemistries, there is a noticeable voltage change as the SoC changes. Therefore, by comparing the measured OCV with a predefined OCV–SoC relationship, it is possible to analyze the SoC of the battery [10,11]. This method can be effective for chemistries with distinct OCV characteristics. Coulomb counting, on the other hand, involves integrating the current flowing in or out of the battery over time. By keeping track of the net charge passing through the battery, it is possible to estimate the SoC. However, accurate current measurement is critical for reliable SoC estimation using this method. Any inaccuracies in current measurement will lead to cumulative errors in SoC estimation over time. Both the OCV measurement and Coulomb counting have their limitations. The OCV method relies on predefined OCV–SoC relationships that may vary due to factors such as temperature and aging. Coulomb counting, while conceptually straightforward, can be prone to cumulative errors caused by inaccuracies in current measurement and drift in capacity due to aging or temperature effects. To improve SoC estimation accuracy, advanced techniques such as model-based methods, Kalman filtering, and adaptive algorithms have been developed. These methods incorporate battery models and utilize additional measurements, such as battery voltage, temperature, and impedance, to enhance SoC estimation accuracy across a wider range of battery chemistries and operating conditions. The accuracy of the Coulomb counting method relies heavily on two factors: precise measurement of battery current and accurate estimation of initial SoC. By integrating the charging and discharging currents over the operating periods and considering a known capacity, SoC can be calculated. However, it's important to note that the releasable charge during the charging and discharging cycles is always less than the stored charge. This is due to losses that occur during the charging and discharging processes, as well as self-discharge of the battery. These losses and self-discharge contribute to accumulating errors in SoC estimation. To achieve more precise SoC estimation, such factors should be taken into account. The losses during charging and discharging, as well as self-discharge, need to be considered and accounted for in the estimation process. Additionally, regular recalibration of the SoC is recommended to maintain accuracy over time. It is also important to consider the decline of the releasable capacity, which refers to the reduction in the amount of charge that can be released by the battery over its lifetime. By incorporating these factors into the estimation method, more accurate SoC values can be obtained.

2.6 MATLAB®-BASED BMS SIMULATION ANALYSIS

The BMS functionality has been simulated using MATLAB Simulink, and the logic has been created using MATLAB State Flow functionality. The parameters used for simulating the BMS are as follows:

- SoC estimation using the Coulomb counting method.
- Cut off the battery above 47°C ambient temperature and below minus 18°C ambient temperature.
- Constant current charging up to 70% of SoC and thereafter constant voltage charging.
- Cut off the battery pack under short circuit current for a specified value above 10 Amps.
- Stop charging rechargeable energy storage system (RESS) at 99% of SoC to avoid overcharging the battery, which could cause irreversible damage.
- Stop discharging RESS below 20% of SoC to avoid permanent damage to the battery and to provide reserve energy, which may be released by intervention from the OEM in case of emergency or to enable the EV to limp back to the nearest service station with degraded performance parameters.

The logic used for implementing these parameters is shown in MATLAB State Flow in Figure 2.3. The MATLAB State Flow functionality is used to generate the required logic for implementing all conditions essential for the effective functioning of the BMS. It permits the design and development of supervisory control, task scheduling, fault management, communication protocols, user interfaces, and hybrid systems.

The following capabilities of Simulink have been utilized for simulating BMS functionality:

- Modeling Battery Packs: Simulink allows the creation of models of battery packs using electrical networks that closely resemble the actual system's topology. These models can be scaled up or down depending on the number of cells in the pack, providing flexibility in system design.
- Parameterizing Equivalent Circuit Elements: To accurately represent the cell chemistry, engineers can parameterize the equivalent circuit elements of the battery model using test data. Simulink facilitates the incorporation of these parameters, enabling a realistic simulation of the battery's behavior.

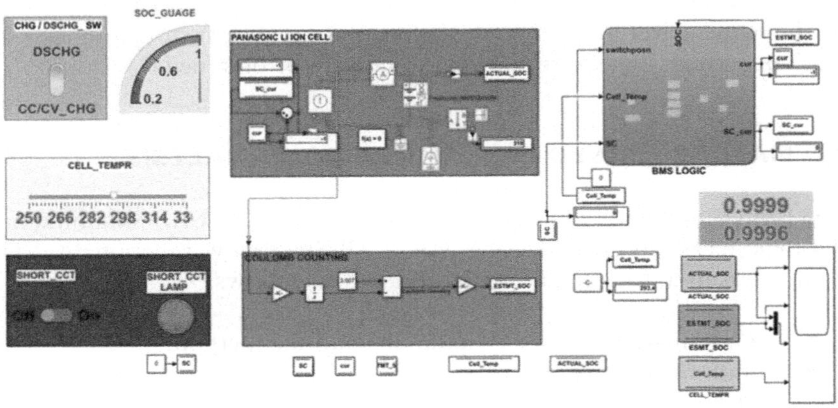

Figure 2.3 MATLAB Simulink BMS model.

- Designing Power Electronics Circuits: Simulink provides tools for outlining the power electronics circuit that connects the battery pack to the control system. It can design and simulate the circuitry to ensure proper integration and functionality.
- Developing Control Algorithms: With Simulink, closed-loop control techniques can be developed for supervisory functions and fault detection logic in the BMS. These algorithms ensure the safe and efficient operation of the battery pack by regulating parameters such as charging, discharging, and temperature.
- Designing State Observers: It supports the design and simulation of state observers for estimating important battery parameters like SoC and SoH in real time. These measurements from the battery system are utilized to estimate the current state and provide valuable information for optimal battery management.
- Verification and Validation: Simulink provides tools for verifying and validating the BMS design through simulation.

Figure 2.3 shows the Simulink model of BMS functionality with Stateflow BMS logic as per the parameters specified above.

The graph in Figure 2.4 shows the actual SoC as determined by the MATLAB Simulink model and the estimated SoC using the Coulomb counting method. Although the Coulomb counting method has a limitation in that it is dependent on the initial SoC value, it is a simple method that does not require significant hardware and software overhead. Figure 2.4 clearly depicts that the actual SoC (as shown by the MATLAB Simulink algorithm) and the estimated SoC based on Coulomb counting are nearly the same, and battery temperature is closely monitored to implement shutdown in case the battery operates outside the permissible operating range.

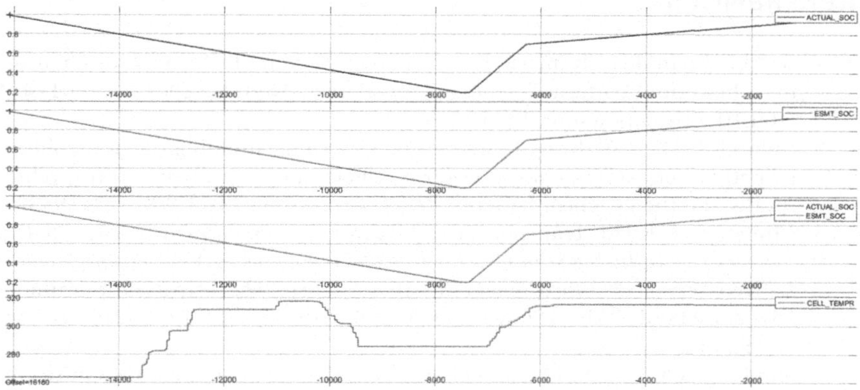

Figure 2.4 Comparison of actual and estimated SoC.

2.7 CONCLUSION

The two-wheeler EV fire incidents are largely attributable to poor quality control of cells during RESS assembly, lack of active fire detection and suppression system, inadequate RESS design, and insufficient BMS functionality. A crucial component of electrical energy storage systems is the BMS. Ensuring the safety and proper functioning of the electrical system relies heavily on a reliable BMS, which effectively responds to both internal and external occurrences. A secure BMS is essential for the successful operation of the electrical system from all angles. In addition to various design parameters discussed, the BMS plays a pivotal role in monitoring and preventing thermal runaway situations. The use of telemetry based on the GSM network can enable real-time warnings to EV users in case of any malfunction or potential thermal runaway, pending corrective actions. The MATLAB-based simulation indicates that reliable BMS functioning can significantly reduce fire incidents in two-wheeler EVs by controlling all aspects of charging and discharging of Li-ion batteries and implementing safety strategies at a lower cost, thereby promoting faster adoption of EVs in the Indian context. BMS-based fire mitigation techniques offer a comprehensive approach to enhancing battery safety. By implementing early detection, fault isolation, thermal management, SoC estimation, communication, and monitoring, along with redundant safety measures, these techniques significantly reduce the risk of battery fires and improve overall system reliability. The field of battery technology and fire mitigation techniques is continuously evolving. BMSs are subject to ongoing research and development to enhance their capabilities in detecting, preventing, and mitigating fire incidents.

REFERENCES

1. Balakrishnan PG, Ramesh R, and Prem Kumar T. "Safety mechanisms in lithium-ion batteries", *Journal of Power Sources* 2006;155:401–414. DOI:10.1016/j.jpowsour.2005.12.002
2. Aalund R, Diao W, Kong L, and Pecht M. "Understanding the non-collision related battery safety risks in electric vehicles a case study in electric vehicle recalls and the LG chem battery", *IEEE Access* June 2021;9:89527–89532.
3. Ministry of Road Transport and Highways. *Amendment 2 to AIS-156: Automotive Industry Standard for Electric Power Train Vehicles of Category L (Two-Wheelers).* Government of India, August 2022. https://hmr.araiindia.com/api/AISFiles/AIS_156_with_Amd2_406992c5-3ef3-499f-a2f4-6677d5c7c7e9.pdf
4. Sun P, Bisschop R, Nui H, and Huang X. "A review of battery fires in electric vehicles", *Research Gate* January 2020;56(4):1361–1410.
5. Justen R, and Schoneburg R., "Crash safety of hybrid- and battery electric vehicles", *22nd International Technical Conference on the Enhanced Safety of Vehicles (ESV),* Washington, DC, 13–16 June 2011.
6. Rakshita R, and Surendra U, "Integration of Coulomb counting method in battery management system for electric vehicle", *International Journal of Engineering Research Technology (IJERT),* May-2020;9(5):946–951.
7. Darwish M, Ionnau S, Janbey A, Amreiz H, and Marouchos CC, "Review of battery management systems", *Proceedings of the International Conference on Electrical, Computer, Communications and Mechatronics Engineering (ICECCME),* Mauritius, 7–8 October 2021.
8. Brandl M, Gall H, Wenger M, Lorentz V, Giegerich M, Baronti F, Fantechi G, Fanucci L, Roncella R, Saletti S, Saponara S, Thaler A, Cifrain M, and Prochazaka W, "Batteries and batteries management system for electric vehicles", *IEEE* 2012;1:1–21.
9. Arora K, Joshi GP, Ragab M, Rawa M, Milyani AH, Mansour RF, and Yang E, "Bilateral contract for load frequency and renewable energy sources using advanced controller", *Computers, Materials & Continua* 2022;73(2):3165–3180, DOI: 10.32604/cmc.2022.026966
10. Bennito A, Rajendran G, Harsha Vardhan S, Ragul Nivash RS, and Balamurugan S, "Real-time implementation of battery management system in electric vehicles", *2021 Innovations in Power and Advanced Computing Technologies (i-PACT)* IEEE, 2021, November 27–29, 2021 in Kuala Lumpur, Malaysia, pp. 1–6.
11. Dorsz A, and Lewandowaski M, "Analysis of fire hazards associated with the operation of electric vehicles in enclosed structures", *MDPI Journal Energies* 2022;15(11):1–13.

Chapter 3

Analyzing solar photovoltaic-supported power systems among diverse external uncertainties

A comprehensive review

Gaurav Kumar, Mansi Gupta, and Krishan Arora

3.1 INTRODUCTION

The integration of solar photovoltaic (PV) systems into power grids has garnered significant attention in recent years, driven by the urgent need to transition toward sustainable energy sources and reduce reliance on fossil fuels [1]. Solar PV technology offers a promising solution for clean and renewable energy generation, with abundant solar resources available worldwide [2]. However, the intermittent nature of solar energy, coupled with various external uncertainties, poses substantial challenges to the seamless integration of solar PV systems into power grids [3]. In this review paper, the authors explore the complexities of incorporating solar PV into power systems amidst external uncertainties such as weather variations, load fluctuations, and grid disturbances [4]. This comprehensive review synthesizes current research findings and examines the methodologies, models, and technologies employed to address these challenges.

The motivation behind integrating solar PV into power systems stems from the environmental imperative to reduce greenhouse gas emissions and mitigate climate change. Solar energy, being abundant, clean, and renewable, presents a viable alternative to conventional fossil fuels [5,6]. However, harnessing solar energy efficiently and reliably requires overcoming numerous technical and operational hurdles, particularly in the presence of external uncertainties. These external uncertainties, including fluctuations in solar irradiance, variations in energy demand, and grid instabilities, can impact the performance, reliability, and stability of solar PV systems [7]. Such uncertainties necessitate robust design strategies and mitigation techniques to ensure the seamless integration of solar PV into power grids. Understanding and addressing these uncertainties is critical for maximizing the benefits of solar energy while maintaining grid stability and reliability.

In this review paper, we explore the design considerations, modeling techniques, mitigation strategies, and experimental investigations related to solar PV-supported power systems under various external uncertainties.

DOI: 10.1201/9781003542568-3

By synthesizing the existing literature and identifying key research gaps, we aim to provide insights into the current state-of-the-art and future directions in this crucial field of research.

3.2 DESIGN CONSIDERATIONS FOR SOLAR PV-SUPPORTED POWER SYSTEMS

Design considerations for solar PV-supported power systems encompass a broad array of factors essential for their effective integration into existing power grids. These considerations include optimizing system topology, sizing components appropriately, and integrating energy storage solutions to mitigate the impact of external uncertainties. Advanced control strategies, such as maximum power point tracking algorithms, play a crucial role in optimizing the performance of solar PV systems [8]. Additionally, the selection of suitable inverters and converters, along with their coordination with grid infrastructure, is paramount to ensuring seamless grid integration and stability [9]. Moreover, the design process must account for dynamic load profiles and weather conditions to accurately assess system performance and reliability [10]. By carefully considering these factors, designers can develop solar PV-supported power systems that are robust, efficient, and capable of withstanding external uncertainties.

3.3 MODELING AND SIMULATION TECHNIQUES

"Modeling and Simulation Techniques" delves into the critical aspect of developing accurate and reliable models for simulating the behavior of solar PV systems under different operating conditions and external uncertainties. Various modeling techniques have been employed to effectively capture the dynamic behavior of solar PV systems. These techniques include solar irradiance modeling, PV module characterization, converter modeling, and system-level simulation frameworks. Solar irradiance modeling plays a fundamental role in predicting the amount of solar energy available for PV generation. Different methods, such as statistical models, satellite data-based models, and numerical weather prediction models, have been used to accurately forecast solar irradiance [11]. Furthermore, PV module characterization involves developing mathematical models to describe the performance characteristics of PV modules under varying environmental conditions. These models consider factors such as temperature, irradiance level, and aging effects to accurately simulate PV module behavior [12]. Converter modeling focuses on capturing the dynamic behavior of power electronic converters used in solar PV systems, such as DC–DC converters and grid-tied inverters. Accurate converter models are essential for designing control strategies and assessing system performance [13]. Additionally, system-level simulation frameworks integrate individual component models to simulate the overall behavior of solar PV systems. These frameworks enable a comprehensive analysis of system performance, including energy yield, grid interaction, and response to external uncertainties [14]. By employing advanced modeling and simulation techniques, researchers can gain valuable insights into the behavior of solar PV systems under various conditions and uncertainties. These insights are crucial for optimizing system design, improving performance, and enhancing grid integration of solar PV-supported power systems.

3.4 MITIGATION STRATEGIES FOR EXTERNAL UNCERTAINTIES

This section addresses the challenges posed by external uncertainties, such as weather variations, load fluctuations, and grid disturbances, in solar PV-supported power systems. Various mitigation strategies have been proposed to enhance the reliability and resilience of these systems. One of the key mitigation strategies involves the development and implementation of forecasting algorithms for solar irradiance and load. Accurate forecasts enable proactive management of solar PV generation and energy demand, reducing the impact of uncertainties on system operation [15]. Advanced control schemes for grid-tied inverters play a crucial role in stabilizing the grid and ensuring seamless integration of solar PV systems. These control schemes include voltage and frequency regulation, active and reactive power control, and anti-islanding protection [16]. Energy storage systems (ESS) serve as another effective mitigation strategy for external uncertainties. ESS can store excess solar energy during periods of high generation and discharge it during periods of low generation or high demand, thus enhancing system flexibility and reliability [17]. Hybridization with other renewable energy sources, such as wind and hydroelectric power, can also mitigate the impact of external uncertainties on solar PV-supported power systems. By diversifying the energy mix, hybrid systems can better cope with fluctuations in renewable energy generation and improve overall system reliability [18].

Furthermore, demand response programs and smart grid technologies enable dynamic adjustment of energy consumption in response to changes in solar PV generation and grid conditions. These demand-side management strategies help balance supply and demand, reducing the need for costly grid upgrades [19]. By implementing these mitigation strategies, solar PV-supported power systems can effectively reduce the impact of external uncertainties, improve reliability, and enhance grid stability.

3.5 CASE STUDIES AND EXPERIMENTAL INVESTIGATIONS

"Case Studies and Experimental Investigations" explores real-world applications and experimental studies conducted to evaluate the performance of solar PV-supported power systems under external uncertainties. One notable case study investigated the impact of weather variations on the performance of a solar PV system installed in a coastal region. The study employed advanced forecasting models to predict solar irradiance and evaluated the effectiveness of different mitigation strategies in improving system reliability and energy yield [20]. Another relevant case study focused on the integration of ESS with solar PV systems to enhance grid stability and reliability. This study involved the deployment of a pilot-scale solar PV-ESS system and assessed its performance under various operating conditions and external uncertainties [21]. In addition to case studies, several experimental investigations have been conducted to evaluate the effectiveness of different design strategies and mitigation techniques for solar PV-supported power systems. For instance, one investigation conducted experiments to assess the performance of advanced control schemes for grid-tied inverters under varying grid conditions and load profiles [22].

Furthermore, field deployments and demonstration projects have provided valuable insights into the practical challenges and opportunities associated with integrating solar PV into power systems. Projects such as the SolarCity microgrid project in California [23] have demonstrated the feasibility and benefits of solar PV-supported power systems in enhancing grid resilience and sustainability. By synthesizing findings from these case studies and experimental investigations, researchers gain valuable insights into the performance, reliability, and resilience of solar PV-supported power systems under various external uncertainties. These insights are essential for informing future design decisions and advancing the integration of solar PV into power grids.

3.6 CHALLENGES AND FUTURE DIRECTIONS

This section highlights key challenges faced in the integration of solar PV systems into power grids amidst external uncertainties and outlines potential future research directions.

- One of the primary challenges is the accurate forecasting of solar irradiance and load demand, which is essential for effective grid operation and energy management [24]. Improving the accuracy of forecasting models and developing robust prediction techniques are crucial areas for future research.
- Another challenge is the limited scalability and flexibility of existing solar PV systems, particularly in urban environments with constrained space and complex shading conditions [25]. Developing innovative design solutions and modular system architectures to overcome these limitations will be essential for expanding the deployment of solar PV systems.
- Grid integration issues, such as voltage fluctuations and power quality concerns, pose significant challenges to the reliable operation

of solar PV-supported power systems [26]. Addressing these challenges requires the development of advanced control strategies and grid-friendly PV inverters to enhance grid stability and resilience.

- Furthermore, the high upfront costs and long payback periods associated with solar PV installations remain a barrier to widespread adoption, especially in developing countries [27]. Exploring innovative financing mechanisms, such as leasing and community solar programs, could help overcome these economic barriers and promote solar PV deployment.
- In terms of future directions, there is a growing need to integrate solar PV systems with emerging technologies such as energy storage, electric vehicles, and smart grid infrastructure [28]. These integrated systems can enhance system flexibility, reliability, and grid support capabilities, paving the way for a more sustainable and resilient energy future.
- Moreover, continued research into advanced materials and manufacturing processes for PV modules and components is essential to improving efficiency, reducing costs, and enhancing the durability of solar PV systems [29].
- The authors delve into the complexities of combining multiple energy sources, such as solar PV, wind, and energy storage, to enhance the efficiency and reliability of renewable energy generation [30]. Exploring novel materials, such as perovskite-based solar cells, holds promise for further advancing the performance and scalability of solar PV technology. Table 3.1 shows the challenges and future directions.

Table 3.1 Challenges and future directions

Challenges	*Future directions*
Accurate forecasting of solar irradiance and load demand	Improve forecasting models and techniques [24]
Scalability and flexibility of solar PV systems in urban environments	Develop innovative design solutions and modular system architectures [25]
Grid integration issues (voltage fluctuations, power quality concerns)	Develop advanced control strategies and grid-friendly PV inverters [26]
High upfront costs and long payback periods for solar PV installations	Explore innovative financing mechanisms such as leasing and community solar programs [27]
Integration with emerging technologies (energy storage, electric vehicles, smart grid infrastructure)	Integrate solar PV systems with energy storage, electric vehicles, and smart grid infrastructure [28]
Continued research into advanced materials and manufacturing processes for PV modules	Research advanced materials and manufacturing processes for PV modules [29]
Challenges include optimizing system efficiency, balancing supply, and demand, and ensuring grid stability	Driving the development and adoption of hybrid energy systems for solar energy [30]

Overall, addressing these challenges and pursuing these future research directions will be crucial for unlocking the full potential of solar PV-supported power systems and accelerating the transition toward a clean and sustainable energy future.

3.7 CONCLUSION

It provides a comprehensive examination of the challenges, design strategies, mitigation techniques, and future directions in the integration of solar PV systems into power grids amidst external uncertainties. Through a synthesis of current research findings and practical insights from case studies and experimental investigations, this review sheds light on the complexities of incorporating solar PV into power systems and identifies key areas for further research and development. The integration of solar PV into power systems offers immense potential for reducing greenhouse gas emissions, enhancing energy security, and promoting sustainable development. However, this transition is not without challenges, as external uncertainties such as weather variations, load fluctuations, and grid disturbances can impact the performance and reliability of solar PV systems. To address these issues, researchers have proposed various design strategies and mitigation techniques, including advanced control schemes, energy storage integration, forecasting algorithms, and hybridization with other renewable energy sources. Additionally, ongoing research efforts focus on improving forecasting accuracy, enhancing grid integration capabilities, reducing costs, and advancing PV technology to unlock the full potential of solar PV-supported power systems. By synthesizing the existing literature and identifying key research gaps, this review paper aims to provide insights into the current state-of-the-art and future directions in this critical field of research. Ultimately, addressing these challenges and advancing solar PV integration will be crucial for accelerating the transition toward a clean, sustainable, and resilient energy future.

REFERENCES

1. Darkwah, W. K., Sandrine, M. K. C., Adormaa, B. B., Teye, G. K., & Puplampu, J. B. (2020). Solar light harvest: modified d-block metals in photocatalysis. *Catalysis Science & Technology*, *10*(16), 5321–5344.
2. Akan, O. B., Ramezani, H., Khan, T., Abbasi, N. A., & Kuscu, M. (2017). Fundamentals of molecular information and communication science. *Proceedings of the IEEE*, *105*(2), 306–318. https://doi.org/10.1109/JPROC.2016.2537306
3. Alotaibi, I., Abido, M. A., Khalid, M., & Savkin, A. V. (2020). A comprehensive review of recent advances in smart grids: a sustainable future with renewable energy resources. *Energies*, *13*(23), 6269.

4. Jaen-Cuellar, A. Y., Elvira-Ortiz, D. A., Osornio-Rios, R. A., & Antonino-Daviu, J. A. (2022). Advances in fault condition monitoring for solar photovoltaic and wind turbine energy generation: a review. *Energies*, *15*(15), 5404.
5. Mekhilef, S., Saidur, R., & Safari, A. (2011). A review on solar energy use in industries. *Renewable and Sustainable Energy Reviews*, *15*(4), 1777–1790.
6. Alonso-Travesset, À., Coppitters, D., Martín, H., & de la Hoz, J. (2023). Economic and regulatory uncertainty in renewable energy system design: a review. *Energies*, *16*(2), 882.
7. Parida, B., Iniyan, S., & Goic, R. (2011). A review of solar photovoltaic technologies. *Renewable and Sustainable Energy Reviews*, *15*(3), 1625–1636.
8. Singh, P., Arora, K., Rathore, U. C., Joshi, G. P., & Cho, W. (2024). Comparative study of controllers in battery energy storage system integrated with doubly fed induction generator-based wind energy conversion system for power quality improvement. *Energy Reports* *11*(9), 4587–4600. https://doi.org/10.1016/j.egyr.2024.04.020
9. Shah, N., Shah, A. A., Leung, P. K., Khan, S., Sun, K., Zhu, X., & Liao, Q. (2023). A review of third generation solar cells. *Processes*, *11*(6), 1852.
10. Lyden, S., & Haque, M. E. (2015). Maximum power point tracking techniques for photovoltaic systems: a comprehensive review and comparative analysis. *Renewable and Sustainable Energy Reviews*, *52*, 1504–1518.
11. Ghosal, M. K. (2022). Entrepreneurship in Renewable Energy Technologies. CRC Press.
12. Yazdanifard, F., Ameri, M., & Taylor, R. A. (2020). Numerical modeling of a concentrated photovoltaic/thermal system which utilizes a PCM and nanofluid spectral splitting. *Energy Conversion and Management*, *215*, 112927.
13. Yusuf, S., & Abubakar, A. (2023). A comprehensive review on grid-forming inverter: potential and future trends. *Majlesi Journal of Electrical Engineering*, *17*(1), 1–27.
14. Lekouaghet, B., Khelifa, M. A., & Boukabou, A. (2022). Adolescent identity search algorithm for parameter extraction in photovoltaic solar cells and modules. *Journal of Computational Electronics*, *21*(4), 859–881.
15. Singh, P., Arora, K., Rathore, U. C., Yang, E., Joshi, G. P., & Son, K. C. (2023). Performance evaluation of grid-connected dfig-based wecs with battery energy storage system under wind alterations using FOPID controller for RSC. *MDPI Journal of Mathematics*, *11*(9), 2100. https://doi.org/10.3390/math11092100
16. Arulkumar, K., Palanisamy, K., & Vijayakumar, D. (2016). Recent advances and control techniques in grid connected PV system: a review. *International Journal of Renewable Energy Research*, *6*(3), 1037–1049.
17. Mądziel, M. (2024). Instantaneous CO_2 emission modelling for a Euro 6 start-stop vehicle based on portable emission measurement system data and artificial intelligence methods. *Environmental Science and Pollution Research*, *31*(5), 6944–6959.
18. Dong, Y. (2022). Topological carbon defects based catalysts for the electrochemical reduction of carbon dioxide and oxygen (Doctoral dissertation, University of British Columbia).
19. Gholizadeh, N., & Musilek, P. (2021). Distributed learning applications in power systems: a review of methods, gaps, and challenges. *Energies*, *14*(12), 3654.

20. Zhang, H. (2016). Impacts of global change on tropospheric ozone and mercury (Doctoral dissertation, Michigan Technological University).
21. Nair, G. B., Nagabhushana, H., Dhoble, N. S., & Dhoble, S. J. (Eds.). (2024). *Emerging Energy Materials: Applications and challenges*. CRC Press.
22. Kedir, N., Nguyen, P. H., Pérez, C., Ponce, P., & Fayek, A. R. (2023). Systematic literature review on fuzzy hybrid methods in photovoltaic solar energy: opportunities, challenges, and guidance for implementation. *Energies*, *16*(9), 3795.
23. Su, X., Zhang, Z., Liu, Y., Fu, Y., Shahnia, F., Zhang, C., & Dong, Z. Y. (2021). Sequential and comprehensive BESSs placement in unbalanced active distribution networks considering the impacts of BESS dual attributes on sensitivity. *IEEE Transactions on Power Systems*, *36*(4), 3453–3464.
24. Arora, K., Joshi, G. P., Ragab, M., Rawa, M., Milyani, A. H., Mansour, R. F., & Yang, E. (2022). Bilateral contract for load frequency and renewable energy sources using advanced controller. *Computers, Materials & Continua*, *73*(2), 3165–3180. https://doi.org/10.32604/cmc.2022.026966
25. Zidane, T. E. K., Aziz, A. S., Zahraoui, Y., Kotb, H., Aboras, K. M., & Jember, Y. B. (2023). Grid-connected solar PV power plants optimization: a review. *IEEE Access*, *11*, 79588–79608.
26. Shafiullah, M., Ahmed, S. D., & Al-Sulaiman, F. A. (2022). Grid integration challenges and solution strategies for solar PV systems: a review. *IEEE Access*, *10*, 52233–52257.
27. Gulagi, A., Choudhary, P., Bogdanov, D., & Breyer, C. (2017). Electricity system based on 100% renewable energy for India and SAARC. *PLoS One*, *12*(7), e0180611.
28. Zhu, J., Li, S., Borghetti, A., Lan, J., Li, H., & Guo, T. (2023). Review of demand-side energy sharing and collective self-consumption schemes in future power systems. iEnergy, 2(2), 119–132.
29. Wu, P., Wang, S., Li, X., & Zhang, F. (2021). Advances in SnO 2-based perovskite solar cells: from preparation to photovoltaic applications. *Journal of Materials Chemistry A*, 9(35), 19554–19588.
30. Falope, T., Lao, L., Hanak, D., & Huo, D. (2024). Hybrid energy system integration and management for solar energy: a review. *Energy Conversion and Management*, *X*, 100527.

Chapter 4

Advanced fault diagnosis and prognostics using digital twin technology

Harpreet Singh Bedi and Tanishk Singhal

4.1 INTRODUCTION

The successful alternative energy resources include solar, wind, biomass, geothermal, and tidal energy, which have emerged as fundamental parts of the shift toward a new energy paradigm in the world. The increasing use of these renewable resources aims to reduce dependency on fossil fuels, address climate change, and enhance energy security. However, the fluctuating and often uncertain nature of these energy sources poses considerable operational and maintenance challenges. Innovations include not only utilizing the capacity of technologies but also enhancing their reliability, efficiency, and scalability.

Non-conventional energy systems often operate within complex environments. For example:

- The offshore wind turbines are exposed to extreme meteorological events, with storms and high humidity specifically raising the probability of mechanical failures.
- Desert-based solar farms often experience losses in efficiency due to the accumulation of dust and high temperatures.
- Corrosion and scaling issues both pose challenges to geothermal systems, affecting the lifespan of infrastructure.

Such complexities call for the generation of innovative solutions that may potentially overcome these issues while simultaneously enhancing performance [1,2]. Under such circumstances, digital twin technology has emerged as a promising innovation with extraordinary capabilities in real-time monitoring, fault diagnosis, and the optimization of non-conventional energy systems.

DOI: 10.1201/9781003542568-4

4.1.1 Traditional maintenance strategies and their limitations

Non-conventional sources of energy have traditionally relied on two main strategies for maintenance:

1. Reactive Maintenance: This method addresses faults only after they occur. While it minimizes initial costs, it tends to result in unplanned downtimes, expensive emergency repairs, and energy losses. For example, a sudden failure of a gearbox in a wind turbine can lead to extended outages and repair costs in the hundreds of thousands of dollars.
2. Preventive Maintenance: Scheduled maintenance aims to avoid risks by conducting regular check-ups and replacing parts [3]. However, this method lacks flexibility, often leading to wasteful servicing, resource wastage, and increased operating costs.

Neither of these approaches addresses the unique needs of alternative energy systems, which require real-time insights along with predictive capabilities to contend with the variabilities associated with renewable resources. These weaknesses of both approaches highlight the need for emergent technologies such as digital twins [4].

4.1.2 Emergence of digital twin technology

Digital twin technology (DTT) is expected to connect the physical system with its virtual replica, thereby enabling a forward-looking strategy for monitoring unconventional energy systems. Digital twins mimic the physics of physical assets in virtual environments using real-time sensor data, advanced analytical methods, and simulation capabilities.

The key novelties that DTT offers are given below:

1. Improved Fault Detection: Real-time fault detection and classification through anomaly-based detection, together with algorithms from machine learning.
2. Prognostics: Accurate estimation of remaining useful life (RUL) for components, allowing operators to plan maintenance activities.
3. Optimization: Different operational scenarios are simulated to enhance energy capture and system efficiency.

In a tidal energy system, for example, digital twins observe and monitor turbine performance in fluctuating water currents, thereby identifying wear on bearings and predicting when such components will need to be replaced [5,6]. DTT analogues can track combustion efficiency in biomass power plants, optimize feedstock utilization, and identify potential equipment malfunctions (Table 4.1).

Table 4.1 Comparison of **maintenance** strategies for non-conventional energy systems

Strategy	*Reactive maintenance*	*Preventive maintenance*	*Digital twin based maintenance*
Fault detection	After failure	Periodic check-ups	Real-time and predictive
Downtime	High	Moderate	Minimal
Cost efficiency	Low	Moderate	Minimal
Sustainability impact	Moderate	Moderate	High

4.1.3 Real-world implications and case studies

The real-world implementations of digital twin technology within alternative energy systems best illustrate its capacity for significant transformation.

1. Offshore Wind Farm Array: Installed in an offshore wind farm, the digital twin system detected abnormal vibration patterns in the gearboxes of three turbines in the installation. The proper intervention avoided major failures, reduced downtime by 35%, and saved $300,000 in repair costs.
2. Desert Region Solar Farm: This ideal DT system enabled the monitoring of continuous panel temperature and detection of dust deposits, improving the cleaning schedule of the 50 MW solar facility. Consequently, optimization led to a 15% increase in annual energy output and a 20% reduction in maintenance costs.
3. Tidal Energy Systems: The tidal power plant used DTT to simulate turbine operation at varying water flow speeds. This helped maintain optimized blade angles, resulting in a 10% improvement in energy generation from the turbine during peak tidal flows.

4.2 FUNDAMENTALS OF DTT

Digital twin technology is, in itself, an innovative strategy for controlling and optimizing intricate systems. DTT creates a virtual model of physical assets in real-time, providing insights, simulations, and predictions that improve both the performance and reliability of a system. This chapter focuses on definitions, fundamental components, architecture, and advantages of DTT, with an emphasis on its application in green energy systems [7].

A digital twin is, essentially, the dynamic virtual representation of an existing physical system or asset that shares real-time data. This two-way synchronization allows the virtual model to replicate the existing conditions of the physical asset, predict its future behavior, and then suggest corrective actions.

The concept of a digital twin was first envisaged in aerospace engineering to enhance aircraft design and performance, as well as to optimize maintenance. The use of DTT is now very widespread in manufacturing, healthcare, and the energy sector. In the renewable energy industry, DTs provide monitoring of wind turbines, solar farms, and energy storage systems for improved performance and risk minimization during operations [8,9].

4.2.1 Physical asset

The physical asset is the physical system replicated by the digital twin. Examples in the green energy sector include wind turbine blades, solar panels, and battery cells. Sensors installed in these assets monitor operational data, including temperature, vibration, and energy output, which form the primary foundation for the operational capabilities of the digital twin [10].

4.2.2 Virtual model

A virtual model can therefore be defined as the computational representation of a physical asset. It uses mathematical algorithms, machine learning frameworks, and simulation tools that replicate the operational characteristics of the actual system [11]. For instance, the digital twin of a wind turbine would simulate rotor dynamics, gearbox performance, and aerodynamic behavior under various wind conditions.

4.2.3 Data integration layer

Data integration is the layer that intervenes between real and virtual items. The result is the smooth transfer of data while allowing the virtual model to be changed immediately without interruptions. Key technologies include

- IoT Sensors: Data capture in the physical system.
- Edge Computing: Processes data locally to reduce latency.
- Cloud Platforms: Store and analyze large datasets for long-term insights [12,13] (Figure 4.1).

4.3 ARCHITECTURE OF DIGITAL TWIN

4.3.1 Data acquisition layer

At this stage, data obtained from implanted sensors in the physical object is summarized. Measured parameters include rotor speed, panel temperature, and battery voltage.

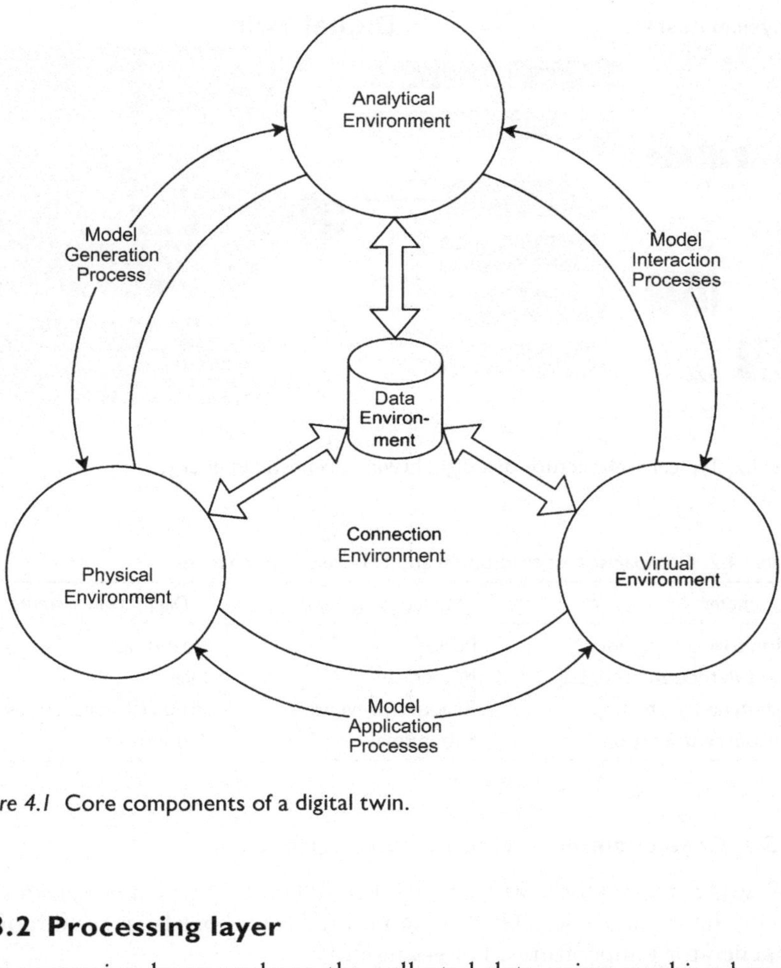

Figure 4.1 Core components of a digital twin.

4.3.2 Processing layer

The processing layer analyzes the collected data using machine learning algorithms, simulation tools, and prediction models. For instance, neural networks (NNs) can detect specific patterns in the vibrations of wind turbines, while finite element models determine the stress distribution in solar panels [14,15].

4.3.3 Interface layer

The interface layer acts as an accessible surface for operator interaction with the digital twin. In this context, it includes dashboards, visualizations, and control systems that present insights, recommendations, and metrics about system health (Figure 4.2, Table 4.2).

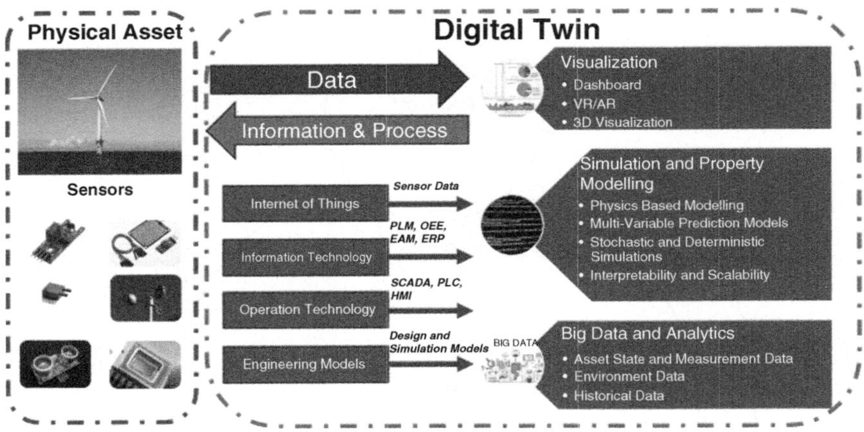

Figure 4.2 Typical architecture of a digital twin in renewable energy.

Table 4.2 Comparison of traditional and DT-based approaches

Parameters	*Traditional systems*	*Digital twin systems*
Monitoring frequency	Periodic	Real time
Fault detection accuracy	Moderate	High
Maintenance strategy	Reactive/preventive	Predictive/prescriptive
Resource utilization	Suboptimal	Optimized

4.3.3.1 Case example – wind turbine digital twin

A DT was deployed for a 50-turbine farm, integrating real-time sensor data with predictive analytics. The DTs monitored rotor speeds, gear vibrations, and generator temperatures. For six months:

- Unplanned downtime decreased by 35%.
- Maintenance costs reduced by $120,000.
- Produced 7% higher energy output with improved blade angles (Figure 4.3).

4.4 DIGITAL TWIN-ENABLED FAULT DIAGNOSIS

A fundamental part of maintaining the reliability and efficiency of green energy systems involves a fault diagnosis procedure, which identifies faults in components such as wind turbine gearboxes, solar inverters, or battery storage units through detection and isolation. Traditional approaches rely on manual inspections or periodic monitoring, which often fail to detect

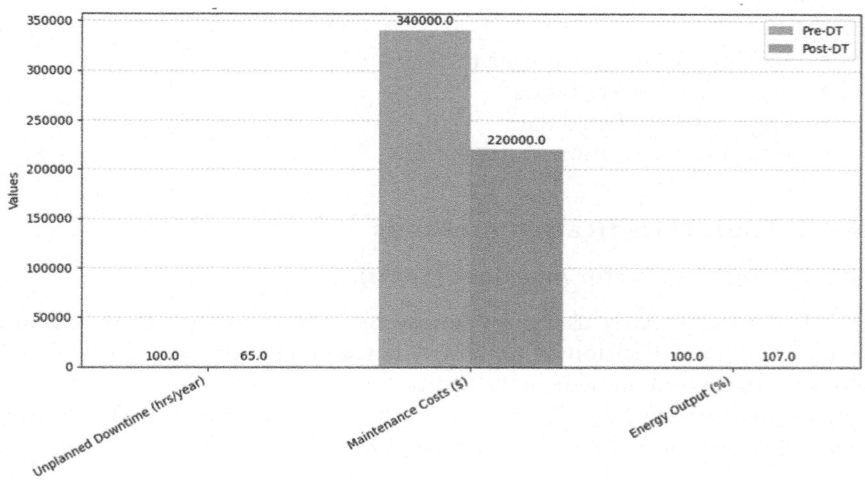

Figure 4.3 Wind turbine performance before and after DT implementation.

slight signs of degradation. DTT transforms this process by allowing real-time, data-informed fault diagnosis based on its capability to simulate, predict, and enhance system performance [16–18].

Real-time monitoring is the foundational framework for fault detection in DTT. Sensors within physical assets continuously collect parameters, including vibration levels, temperatures, and energy consumption. These parameters are transmitted to a digital twin, which analyzes them against expected norms [19].

4.4.1 Statistical methods

Anomalies can be identified by studying historical patterns using statistical frameworks, such as moving average and principal component analysis, among others.

For example, an abnormal fluctuation in the temperature of a wind turbine gearbox may indicate a lubrication fault.

4.4.2 Machine learning models

Machine learning models, particularly autoencoders and NNs, can effectively detect anomalies. They learn the normal operational behavior of systems and are trained to recognize data points that significantly deviate from that pattern.

Equation 4.1: Anomaly Score

$$A(t) = \frac{\left|X_{\text{observed}}(t) - X_{\text{predicted}}(t)\right|}{\sigma} \tag{4.1}$$

where

- $A(t)$: anomaly score at time t
- $X_{\text{observed}}(t)$: observed data
- $X_{\text{predicted}}(t)$: predicted data
- σ: standard deviation of predictions

4.4.3 Fault classification methods

4.4.3.1 Support vector machines (SVMs)

SVMs are particularly useful for binary or multi-class fault classification. They determine decision boundaries that can separate the normal and faulty states based on their input features.

Example: An SVM can be used to classify rotor imbalances, bearing wear, or overheating in a wind turbine [20].

4.4.3.2 Neural networks

These types of NNs, particularly convolutional NNs, are extremely efficient for pattern recognition in high-dimensional datasets.

Example: A CNN trained on vibration data may be able to detect faults such as cracked turbine blades or misaligned bearings.

4.4.3.3 Decision trees (DTs)

A DT creates a hierarchical classification of faults based on input features. This approach is simple yet quite effective for systems where fault categories are well-defined (Figure 4.4, Table 4.3).

4.4.3.3.1 Case example – fault diagnosis in wind turbines

A wind farm applied DTT throughout its entire array of turbines. The main results consisted of

- Anomaly Detection: Early signals of wear in the bearings of 5 turbines.
- Fault Classification: The DT classified these as Stage-2 bearing faults using NNs.
- Optimization of Maintenance: Scheduled low-season maintenance resulted in reduced downtime by 35%, saving $200,000 in repair costs (Table 4.4).

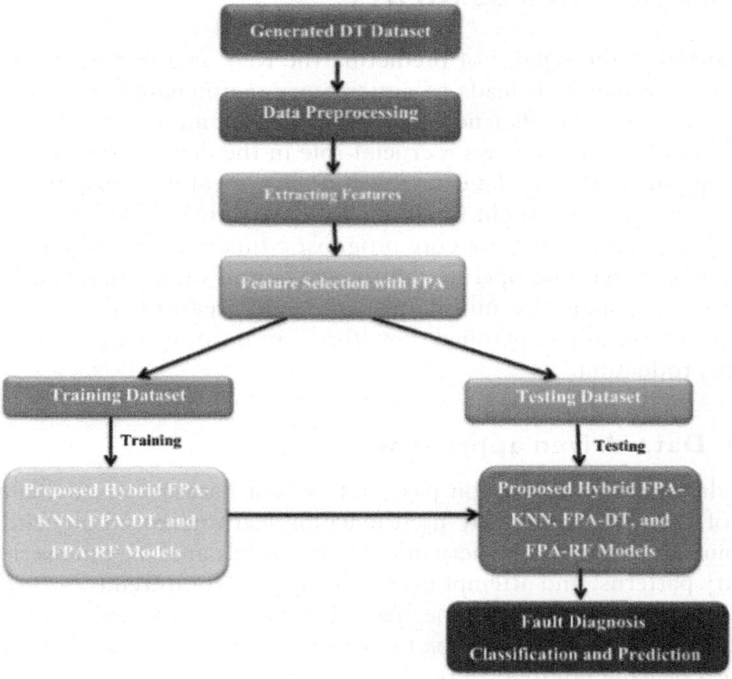

Figure 4.4 Fault diagnosis workflow in digital twin technology.

Table 4.3 Fault classification accuracy by model

Model	*Accuracy (%)*	*Precision (%)*	*Recall (%)*	*F1 score (%)*
Neural networks	96.5	97.0	96.2	96.8
SVM	93.4	94.1	92.8	93.4
Decision trees	88.3	87.9	88.5	88.1

Table 4.4 Fault diagnosis efficiency pre-DT and post-DT

Metric	*Pre-DT*	*Post-DT*	*Improvement (%)*
Fault detection accuracy	85%	96%	12
Downtime (hours/year)	200	130	35
Maintenance cost ($)	250,000	200,000	20

4.5 PROGNOSTICS USING DTT

Prognostics is the science of predicting the RUL and potential failures of system components. It leads to anticipatory maintenance, reduced downtime, and increased efficiency of the overall system, as it predicts when a fault may occur. DTT plays a crucial role in the development of prognostics, applying real-time data, establishing historical patterns, and simulating how components might function at different times [21–23].

RUL estimation is also a core prognostic function, providing clear predictions on remaining operational time. The end of a component's operational life is usually examined through the lifespan of renewable energy systems, where any unplanned downtime can result in significant losses in energy production.

4.5.1 Data-driven approaches

Data-driven methods rely on past and present datasets to calculate RUL. One of the most commonly used machine learning-based algorithms for this purpose is the LSTM network. These models analyze time-series data, identify patterns, and attempt to forecast degradation trends.

Example: In a wind turbine, LSTM networks can analyze data from vibration sensors in the gearbox to estimate the remaining operational time before a bearing failure occurs.

4.5.2 Physics-based models

Physics-based models simulate component degradation mechanisms, such as wear, fatigue, and corrosion. First-principle equations are employed to predict RUL based on available operating conditions and material properties [24].

Example: A physics-based model can simulate how cracks propagate in turbine blades under time-varying wind loads, accurately predicting when a blade will need to be replaced.

4.5.3 Hybrid techniques

Hybrid methodologies integrate data-centric and physics-informed strategies, leveraging the strengths of both approaches to enhance prediction accuracy. In solar energy installations, hybrid models combine historical data on panel efficiency with outcomes from thermal simulations to forecast potential inverter failures [25] (Figure 4.5).

Equation 4.2: RUL Estimation Formula

$$\text{RUL} = T_{\text{failure}} - T_{\text{current}} \tag{4.2}$$

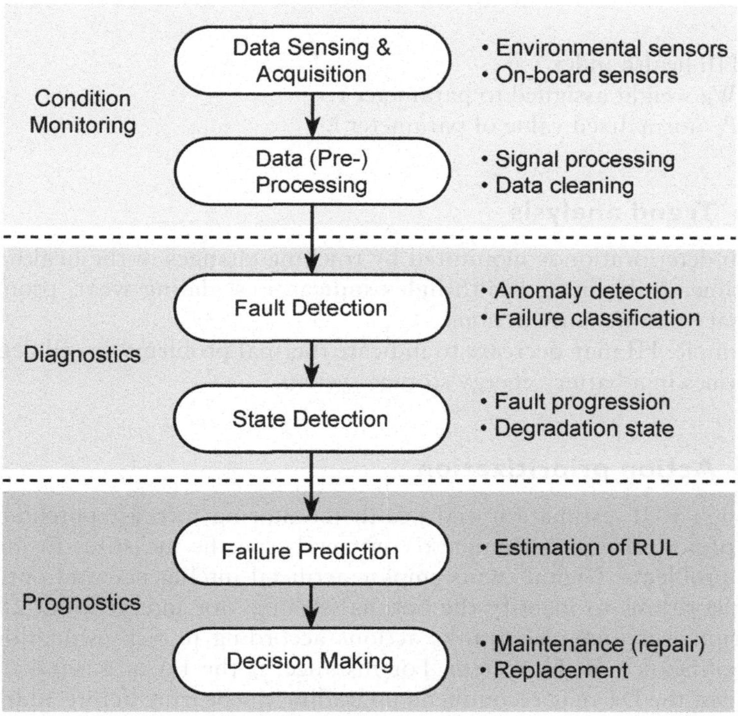

Figure 4.5 RUL prediction workflow in digital twin technology.

where

- $T_{failure}$: predicted failure time
- $T_{current}$: current operational time

4.5.4 Aggregation of multi-sensor data

A Health Index gives an overview of the health of a system. This index combines values such as temperature, vibration, and performance parameters into a single numerical value. The calculation of health indices is done by integrating sensor readings using weighted averages or machine learning algorithms [26,27].

Equation 4.3: Health Index Calculation

$$\mathrm{HI} = \sum_{i=1}^{n} w_i \cdot P_i \tag{4.3}$$

where

- HI: health index
- W_i: weight assigned to parameter i
- P_i: normalized value of parameter i.

4.5.5 Trend analysis

System deterioration is monitored by tracking changes in the health index over time. A declining health index indicates escalating wear, prompting necessary maintenance actions.

Example: HI may decrease to indicate thermal problems or cell degradation issues in a battery energy storage system.

4.5.6 Action prioritization

Although RUL estimation and health monitoring forecast potential failures, prescriptive maintenance recommends specific measures to mitigate likely problems. Digital twins employ artificial intelligence and optimization algorithms to identify the optimal strategy for intervention [28–30]. Prescriptive maintenance ranks actions according to risk evaluation and the significance of the system. For instance, if the HI of a wind turbine degrades, the DT may recommend upgrading the bearing before addressing less pressing rotor balance issues.

4.5.7 Maintenance resource optimization

Digital twins allow two enhanced resources to be allocated: spare components, workforce allocation, and maintenance schedules. A solar farm diagnostic tool can even recommend cleaning only those panels that have experienced significant efficiency degradation, thereby reducing operational expenses (Table 4.5).

Table 4.5 Prescriptive maintenance recommendations

System component	*Detected issue*	*Recommended action*	*Priority*
Wind turbine gearbox	Vibration anomaly	Replace bearings	High
Solar panel string	Shading detected	Clean panels	Medium
Battery storage module	Temperature deviation	Adjust thermal management	High

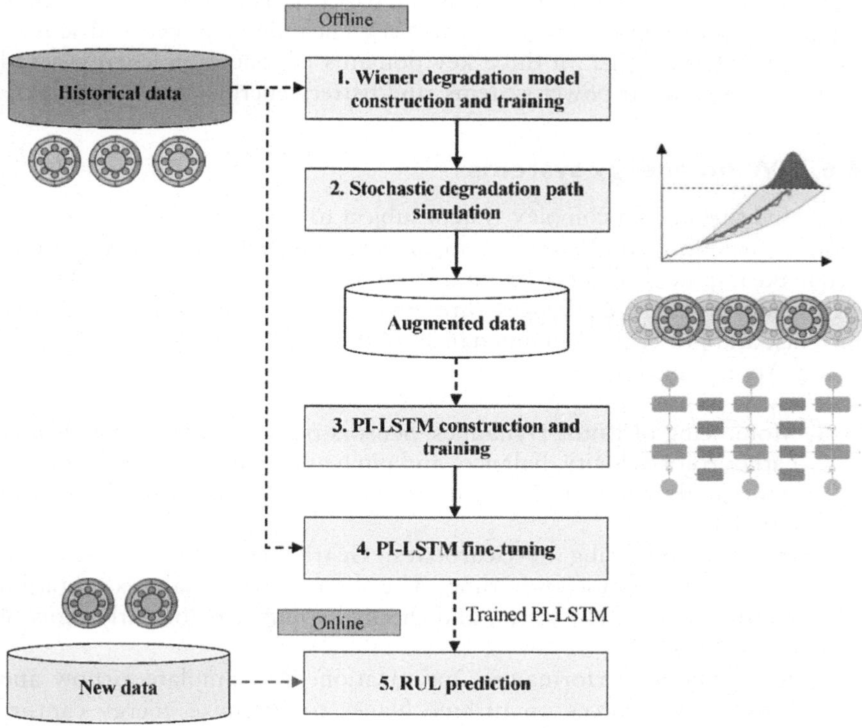

Figure 4.6 RUL predictions for wind turbine gearboxes.

4.5.7.1 Case example – prognostics in wind turbines

Utilizing this opportunity, one wind farm employed DTT to monitor and predict the condition of its turbines in operation. The digital twins used a hybrid prognosis framework incorporating LSTM networks and finite element analysis. The main results included (Figure 4.6):

- RUL Prediction: Gearbox failure was predicted 30 days in advance, allowing time for scheduled maintenance.
- Cost Savings: Early intervention reduced repair costs by 40%.
- Increased Reliability: Turbine availability increased by 15%.

4.6 APPLICATIONS OF DTT IN GREEN ENERGY

DTT is a novel instrument in most renewable energy domains. This technology enables highly accurate virtual representations of tangible assets, allowing for a comprehensive understanding of system functionality to facilitate decisions on anticipatory maintenance and enhancing operational capabilities while optimizing resources [31]. The following section discusses the application of DTT in three key domains of renewable energy: wind energy systems, solar power systems, and battery energy storage systems.

4.6.1 Wind energy systems

The wind turbine is a complex system subject to variations in external conditions, including wind speeds, temperature, and humidity. These factors can make it quite difficult to maintain and optimize the system [32]. Digital twins address these challenges through real-time monitoring, predictive insights, and prescriptive maintenance solutions.

Key Applications:

1. Monitoring of Rotor Dynamics: Sensors placed in the turbine blades gather rotor velocity, balance, and pitch angle data, which the digital twins analyze to identify any imbalances that might cause vibro-shake damage.
2. Health Monitoring of Gearboxes: Gearboxes are critical components that degrade over time. The digital twin analyzes vibration metrics, lubrication states, and thermal conditions for early signs of degradation.
3. Aerodynamic Performance Optimization: DTs simulate airflow and aerodynamic forces on turbine blades to optimize energy capture under varying wind conditions (Figure 4.7, Table 4.6).

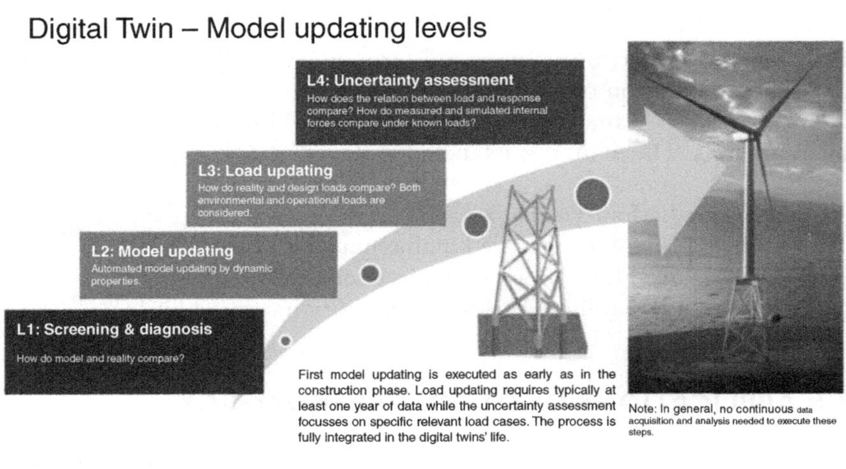

Figure 4.7 Digital twin application in wind turbine rotor dynamics.

Table 4.6 Wind energy metrics pre-DT and post-DT implementation

Metric	*Pre-DT*	*Post-DT*	*Improvement (%)*
Downtime (hours/year)	150	100	33
Maintenance cost ($)	120,000	90,000	25
Energy output (MWh/year)	8,500	9,180	8

Table 4.7 Solar farm improvements using DTT

Metric	*Pre-DT*	*Post-DT*	*Improvement (%)*
Energy loss (%)	10%	6%	40
Downtime (hours/year)	50	30	40
Maintenance cost ($/year)	50,000	35,000	30

4.6.2 Solar power systems

Solar farms encounter distinct challenges, including the buildup of dust, shading effects, and inverter failures. The implementation of digital twins facilitates real-time observation and analysis, enhancing system performance and minimizing energy losses [33,34].

Key Applications:

1. Panel Efficiency Diagnostics: DTs monitor temperatures of panels, current flow, and output voltage to detect inefficiencies caused by shading, dirt, or damage.
2. Inverter Monitoring and Optimization: Inverters convert DC power to AC. The DTs analyze the performance of inverters to detect overheating, aging, or otherwise inefficient conditions.
3. String-Level Monitoring: DTs monitor the performance of every panel string to balance the power generated by the array (Figure 4.8, Table 4.7).

4.6.3 Battery energy storage systems

If the renewable energy supply is not stabilized, its reliability may be compromised. Additionally, batteries experience degradation due to cycling, thermal stress, and aging. Enhanced digital twins improve battery performance and longevity through prognostic insights and resource optimizations [35–37].

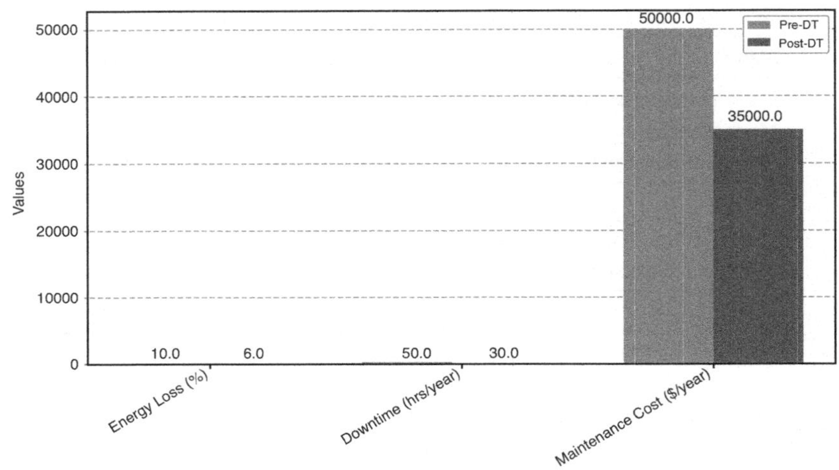

Figure 4.8 Solar farm performance optimization using digital twins.

Table 4.8 Battery storage performance metrics pre-DT and post-DT

Metric	*Pre-DT*	*Post-DT*	*Improvement (%)*
Battery life (cycles/year)	800	920	15
Thermal issues (events/year)	10	3	70
Capacity utilization (%)	85%	95%	12

Primary Applications:

1. Battery Life Management: The DTs predict cell degradation and optimize charge–discharge cycles to extend the life cycle.
2. Thermal Control: DTs regulate cell temperatures and mimic thermal circulation to avoid overheating.
3. Capacity Forecasting: DTs assess the available storage capacity under various load scenarios, facilitating improved energy management (Figure 4.9, Table 4.8).

4.7 RESULTS AND DISCUSSIONS

The incorporation of DTT into green energy systems has led to significant improvements in operational efficiency and accuracy in fault detection, while also providing cost savings. This section will discuss the results presented due to the use of DTT in different renewable energy sectors, along with data, tables, and graphic representations.

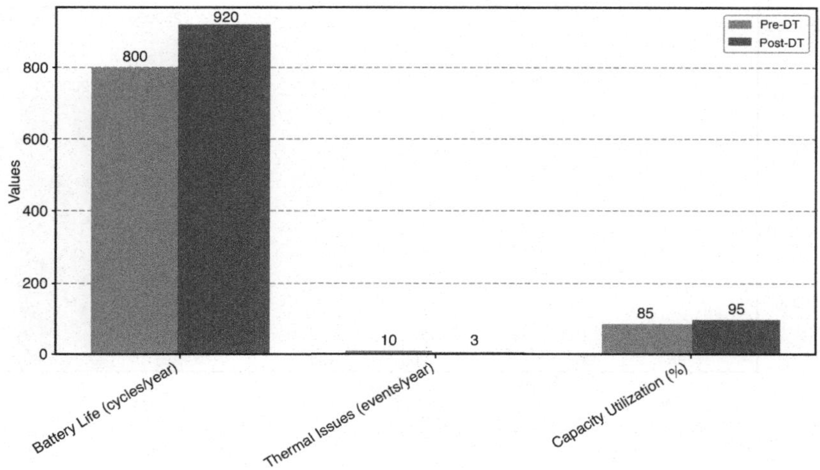

Figure 4.9 Battery performance monitoring using digital twins.

4.7.1 Enhanced fault detection efficiency

Fault detection is critical for minimizing downtime and ensuring system reliability. Digital twins have been shown to significantly improve fault detection over traditional approaches [38–40].

Case Study of Wind Turbine Fault Detection: In a 50-turbine wind farm, digital twins were applied to control and monitor gearbox vibrations and rotor dynamics. These digital twins detected initial faults in 7 turbines, allowing for advanced maintenance planning (Figure 4.10, Table 4.9).

4.7.2 Improved energy output

Digital twins optimize performance by identifying inefficiencies and suggesting operational adjustments.

Case Study: Optimization of a 10 MW solar farm assisted by DTT for monitoring panel efficiency and inverter performance. Over a period of more than 12 months, the DT was able to detect shading problems, inverter degradation, and panel hotspots, which resulted in substantial gains in energy output (Figure 4.11, Table 4.10).

4.7.3 Cost savings in maintenance

The forecasting and advisory functions of digital twins enable operators to plan maintenance activities effectively in order to prevent unnecessary interventions and expensive repairs.

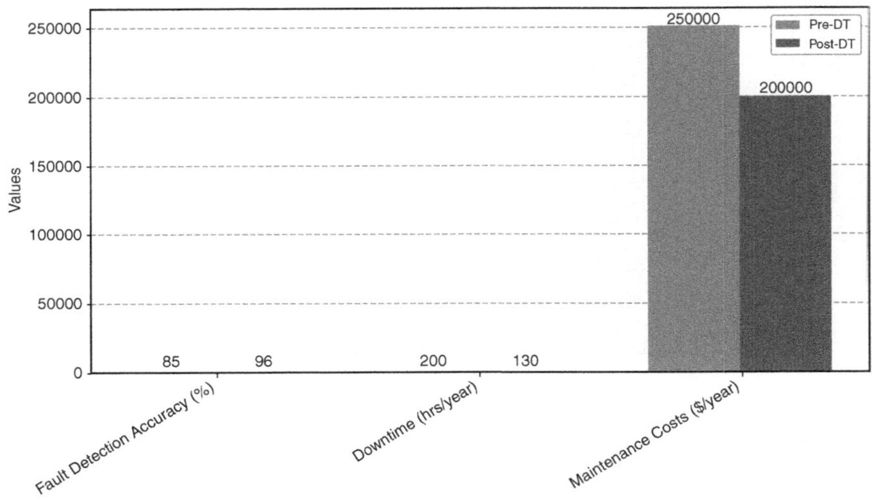

Figure 4.10 Comparative analysis of fault detection accuracy.

Table 4.9 Fault detection metrics pre-DT and post-DT

Metric	*Pre-DT*	*Post-DT*	*Improvement (%)*
Fault detection accuracy	85%	96%	12
Downtime (hours/year)	200	130	35
Maintenance costs ($/year)	250,000	200,000	20

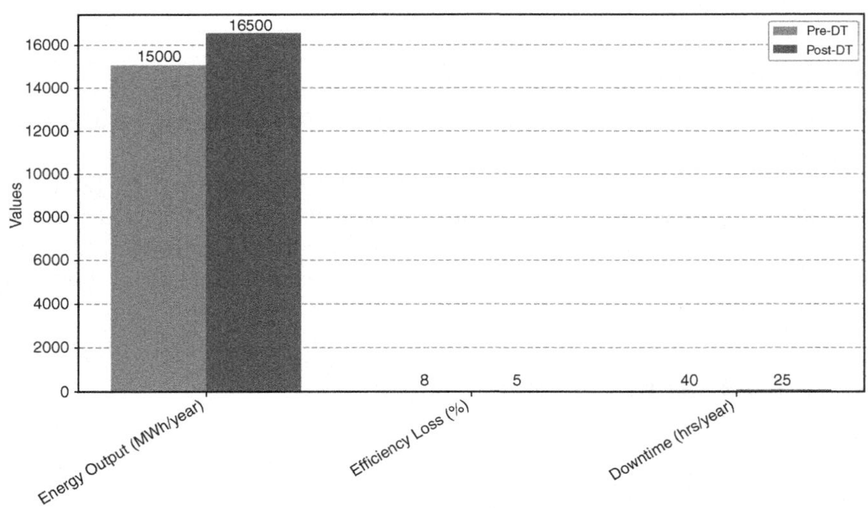

Figure 4.11 Solar farm energy output before and after DTT implementation.

Table 4.10 Solar farm energy output metrics

Metric	*Pre-DT*	*Post-DT*	*Improvement (%)*
Energy output (MWh/year)	15,000	16,500	10
Efficiency loss (%)	8%	5%	37.5
Downtime (hours/year)	40	25	37.5

Case study of Battery Storage Maintenance: A battery energy storage system utilized DTs to monitor cell health and thermal behavior. By predicting cell degradation and optimizing cooling operations, the DT significantly reduced maintenance costs (Figure 4.12, Table 4.11).

4.7.4 Reduction in downtime

Digital twins reduce downtime by enabling early fault detection and optimal maintenance scheduling.

Case Study of Reducing Downtime in Wind Turbines: A set of 2 MW commercial offshore wind turbines applied diagnostic technologies to measure gearbox integrity and rotor dynamics. Maintenance work was scheduled when energy demand was reduced which led to maximum minimization of standby time (Figure 4.13, Table 4.12).

4.7.5 Health index monitoring insights

Digital twins provide a broad Health Index as a representation of the status of assets. Continuous monitoring of the HI ensures timely interventions.

Case Study Health Index Trends in Solar Panels At a solar farm, the DT followed panel HI trends over 12 months. The panels with decreasing HI were cleaned or replaced to maintain efficiency at its best (Figure 4.14).

4.8 CHALLENGES AND FUTURE DIRECTIONS

DTT has shown great promise in improving performance and reliability in green energy systems, there are still some issues with its application. These relate to technical, operational, and cybersecurity aspects. Efforts are important here to overcome these issues, as scaling up DTT across renewable energy will be crucial. The section that follows discusses directions forward, specifically opportunities for further improving DTT toward the optimization of green energy systems.

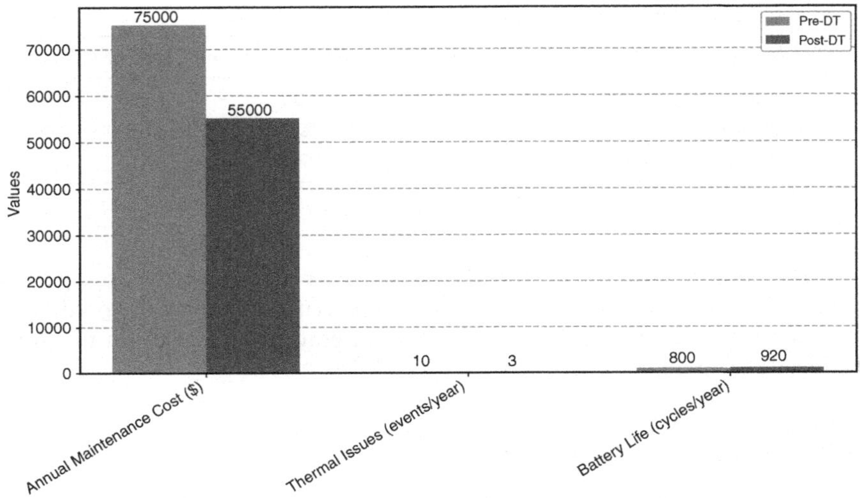

Figure 4.12 Annual maintenance costs pre-DT and post-DT.

Table 4.11 Maintenance cost comparison

Metric	*Pre-DT*	*Post-DT*	*Improvement (%)*
Annual maintenance Cost ($)	75,000	55,000	26.7
Thermal issues (events/year)	10	3	70
Battery life (cycles/year)	800	920	15

4.8.1 Data integration from heterogeneous sources

Digital twins are data-intensive and sourced from a variety of sources: from IoT sensors to meteorological forecasts to operational records. The challenge in integrating all these different streams in real time is quite high:

1. Data Inconsistency: Variations in format, sampling rate, and communication protocols in the data make integration challenging.
2. Legacy Systems: Renewable energy installations, particularly the earlier generations of wind turbines and solar farms, rely on legacy systems that are incompatible with newer data platforms.
3. Scalability: The task becomes computationally intensive at scale as more and more assets are monitored.

Implementing open data standards such as OPC-UA and utilizing cloud and edge computing will allow for effective management of extensive data processing.

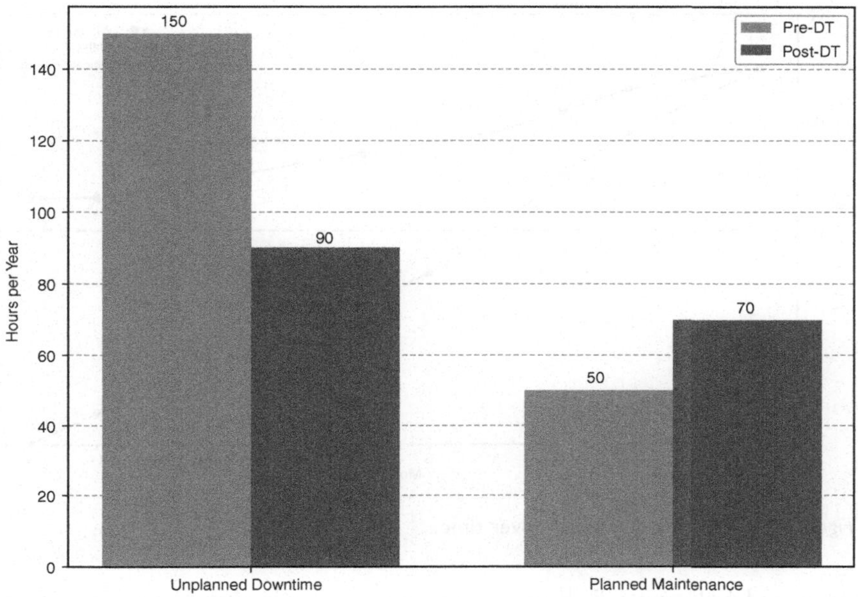

Figure 4.13 Downtime before and after digital twin integration.

Table 4.12 Downtime reduction metrics

Metric	*Pre-DT*	*Post-DT*	*Improvement (%)*
Unplanned downtime (hours/year)	150	90	40%
Planned maintenance hours	50	70	40% (shifted)

4.8.2 Cyber-security risks

Digital twins, requiring constant connectivity as well as data exchange, are vulnerable to cyber-attacks. The fundamental cybersecurity weaknesses include

1. Data Breach: Unauthorized access to sensitive operational data may compromise system integrity.
2. Control Manipulation: Hackers could potentially manipulate digital twins to disrupt system operations.
3. IoT Vulnerabilities: The sensors and devices used in DTT are likely the weakest links in the cybersecurity framework.

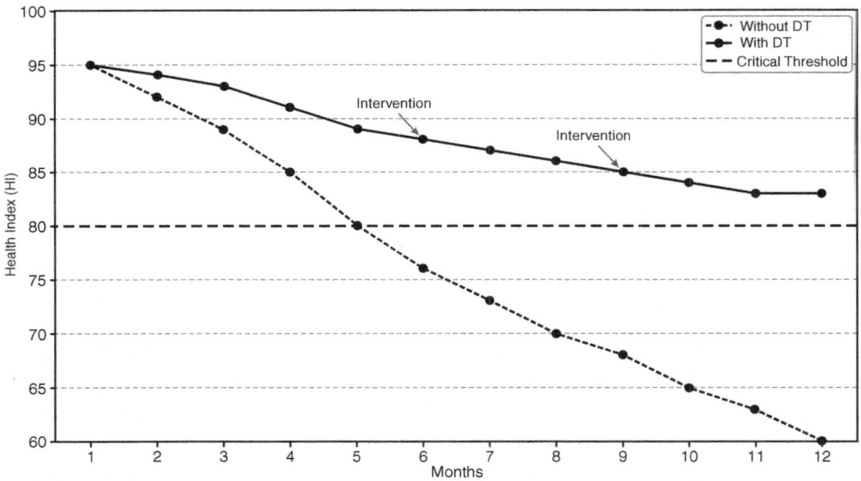

Figure 4.14 Health index trends over time.

Proposed Solutions: Robust encryption standards should be used in the data transmission process. Blockchain technology can secure data integrity. IoT firmware should be frequently updated to patch known vulnerabilities.

4.8.3 Model calibration and scalability

Digital twin models are required to precisely mirror the physical systems they depict. This necessity entails continuous calibration, especially as these systems mature or experience alterations.

1. Model Drift: Changes occur between the digital twin and the real physical system either because of deterioration of components or changes in operational conditions.
2. Scalability Challenges: The scalability of digital twin applications to large sets of wind turbines or solar panels poses significant computational challenges.

Automated machine learning methodologies were employed to develop models efficiently, as well as distributed computing frameworks to enhance scalability.

4.8.4 AI-enhanced digital twins

Artificial intelligence can significantly augment the functionality of digital twins by permitting:

1. Adaptive Learning: AI algorithms allow digital twins to adapt to changing system conditions, improving their predictive accuracy.

2. Self-Decision-Making Power: AI-based DTs can autonomously decide to apply corrective measures, such as adjusting wind turbine pitch angles or realigning the solar panel angle.
3. Enhanced Fault Prediction: Advanced AI models, such as generative adversarial networks, can simulate fault scenarios for more accurate predictions.

Example: An AI-augmented digital twin can predict thermal runaway in a battery energy storage system 95% of the time, facilitating mitigation through preventive cooling.

4.8.5 Self-healing systems with robotic integration

In addition, they could provide robotics for automated maintenance and repair, as well as self-healing energy systems.

1. Robotic Inspection: Using autonomous drones or robots equipped with cameras and sensors, wind turbine blades or solar panels could be inspected for damage.
2. Self-Repairing Robots: Such devices could seal tiny cracks at the root of turbine blade defects, clean solar panels, and perform similar tasks.

In offshore wind installations, automated robotic systems may independently perform repairs on nacelle components, thereby reducing reliance on expensive human-operated maintenance vessels.

4.8.6 Industry-wide standardization

The lack of standardized frameworks for digital twin implementation hinders adoption. Developing industry-wide standards would:

1. Interoperability: Advance seamless integration of digital twins with various systems and manufacturers.
2. Encourage Cooperation: Facilitate data sharing and best practices within the renewable energy community.
3. Accelerate Adoption: Simplify deployment processes, making DTT accessible to smaller operators.

Organizations such as the International Electrotechnical Commission and the Digital Twin Consortium are currently working on developing the standards.

4.8.7 Quantum computing for digital twins

Quantum computing offers high efficiency in simulating complex systems with high computational intensity and has the potential to revolutionize digital twin simulations.

1. Faster Simulations: Quantum algorithms enable quicker simulations of turbine aerodynamics or energy flow in solar farms.
2. Higher Prediction Accuracy: Quantum machine learning algorithms may provide greater prediction accuracy in fault detection due to the parallel analysis of large datasets.

A real-time, quantum-enhanced digital twin of a wind farm could fine-tune blade designs to increase captured energy levels by up to 15%.

4.9 CONCLUSIONS

DT is gradually evolving as the next-generation solution to address the challenges of green energy systems. DTT bridges the gap between physical assets and their digital counterparts, making it an excellent approach for real-time monitoring, fault detection, predictive analytics, and operational enhancement. This chapter outlines the fundamental concepts, implementation, challenges, and prospects of DTT, presenting this technology in its transformative role within renewable energy systems.

REFERENCES

1. Dalibor M, Jansen N, Rumpe B, Schmalzing D, Wachtmeister L, Wimmer M and Wortmann A. A cross-domain systematic mapping study on software engineering for digital twins. *J Syst Softw* 2022;193:111361.
2. Gui Y, Nainar K, Ciontea CI, Bendtsen JD, Iov F, Shahid K and Stoustrup J. Automatic voltage regulation application for PV inverters in lowvoltage distribution grids–a digital twin approach. *Int J Electr Power Energy Syst* 2023;149:109022.
3. Hang T. Digital twins of multiple energy networks based on real-time simulation using holomorphic embedding method, part II: data-driven simulation. *Int J Electr Power Energy Syst* 2023;153:109325.
4. Moutis P and Alizadeh-Mousavi O. Digital twin of distribution power transformer for real-time monitoring of medium voltage from low voltage measurements. *IEEE Trans Power Deliv* 2020;36(4):1952–1963.
5. Gauce D, Lektauers A, Solovjova I, Grants R, Kolosovs D and Litvinenko A. Application of digital twin in medium-voltage overhead distribution network inspection. *Remote Sens* 2023;15(2):489.
6. Mateo Domingo C, San Roman TG, Sanchez-Miralles A, Gonzalez JPP and Martinez AC. A reference network model for large-scale distribution planning with automatic street map generation. *IEEE Trans Power Syst* 2011;26(1):190–197.
7. Abur A, and Gómez-Expósito A. *Power System State Estimation: Theory and Implementation*. CRC Press; 2004.
8. Primadianto A, and Lu C. A review on distribution system state estimation. *IEEE Trans Power Syst* 2017;32(5):3875–3883.

9. Verma P, Bedi H, and Sharma RK. Integrated utilization of solar energy in a zero energy building (ZEB): an approach towards sustainable development. *IJCTA*, 2016;9(41):435–441.
10. Malik P, Sneha, Garg D, Bedi H, Gehlot A and Malik PK. An improved agriculture farming through the role of digital twin. In *2023 4th International Conference on Electronics and Sustainable Communication Systems (ICESC), Coimbatore, India*, 2023, pp. 667–671. http://doi.org/10.1109/ICESC57686.2023.10193522
11. Mwango M, Shrey Y, Bedi HS and Dhillon J. PV system design and solar generation implementation. In: Khosla A, and Kolhe M. (eds), *Renewable Energy Optimization, Planning and Control. Studies in Infrastructure and Control.* Springer, Singapore, 2023.
12. Geth F, Vanin M, and Van Hertem D. Data quality challenges in existing distribution network datasets. In: *Proceedings of the CIRED Conference*, Rome, Italy, 2023, p. 1–6.
13. Lave M, Reno MJ, and Peppanen J. Distribution system parameter and topology estimation applied to resolve low-voltage circuits on three real distribution feeders. *IEEE Trans Sustain Energy* 2019;10(3):1585–1592.
14. Vanin M, Geth F, D'hulst R, and Van Hertem D. Combined unbalanced distribution system state and line impedance matrix estimation. *Int J Electr Power Energy Syst* 2023;151:109155.
15. Bedi HS, Malik PK, Singh R, Singh R, and Bisht N. Traffic surveillance using computer vision and deep learning methods." *IOP ConfSer Earth Environ Sci* 2024;1285(1):1–14.
16. Rigoni V et al. Representative residential LV feeders: a case study for the North West of England. *IEEE Trans Power Syst* 2016;31(1):348–360.
17. Zhang Z, Li R, and Li F. A novel peer-to-peer local electricity market for joint trading of energy and uncertainty. *IEEE Trans Smart Grid* 2019;11(2):1205–1215.
18. Groß D, Früh H, Wiest P, Contreras D, Rudion K, Rupp L, and Lakenbrink C. Evaluation of a three-phase distribution system state estimation for operational use in a real medium voltage grid. In: *IEEE PES ISGT Europe*, IEEE, 2019, p. 1–5.
19. Früh H, Rudion K, von Haken A, Groß D, and Wasowicz B. Evaluation of a three-phase distribution system state estimation for operational use in a real low voltage grid. In: *Proceedings of the 9th Renewable Power Generation Conference*. Online, 2021, p. 125–130.
20. Koirala A, Suárez-Ramón L, Mohamed B, and Arboleya P. Non-synthetic European low voltage test system. *Int J Electr Power Energy Syst* 2020;118:105712.
21. Sandell S, Bjerkehagen D, Birkeland B, and Sperstad IB. Dataset for a Norwegian medium and low voltage power distribution system with industrial loads. *Data Brief* 2023;48:109121.
22. Fan Z, Cao J, Jamal T, Fogwill C, Samende C, Robinson Z, Polack F, Ormerod M, George S, Peacock A, and Healey D The role of 'living laboratories' in accelerating the energy system decarbonization. *Energy Rep* 2022;8:11 858–864.
23. Bedi H. S, Rai R, Viresh S, Kumar S, Kumar A, and Adarsh. Design and implementation of wireless multiple agriculture robot. *J NanoElectron Phys* 2024;16(4):Article 04009. https://doi.org/10.21272/jnep.16(4).04009

24. Nanchian S, Majumdar A, and Pal BC. Ordinal optimization technique for three-phase distribution network state estimation including discrete variables. *IEEE Trans Sustain Energy* 2017;8(4):1528–1535.
25. Negri E, Fumagalli L, and Macchi M. A review of the roles of digital twin in CPS-based production systems. *Proc Manuf* 2017;11:939–948.
26. Arora K, Kumar A, Kamboj VK, Prashar D, Shrestha B, and Joshi GP. Impact of renewable energy sources into multi area multi-source load frequency control of interrelated power system. *MDPI J Math* 2021;9(2):186. https://doi.org/10.3390/math9020186
27. Schleich B, Anwer N, Mathieu L, and Wartzack S. Shaping the digital twin for design and production engineering. *CIRP Ann* 2017;66(1):141–144.
28. Schmidt B and Wang L. Cloud-enhanced predictive maintenance. *Int J Adv Manuf Technol* 2016;99(1–4):5–13.
29. Söderberg R, Wärmefjord K, Carlson JS, and Lindkvist L. Toward a digital twin for real-time geometry assurance in individualized production. *CIRP Ann* 2017;66(1):137–140.
30. Tao F, Cheng J, Qi Q, Zhang M, Zhang H, and Sui F. Digital twin-driven product design, manufacturing and service with big data. *Int J Adv Manuf Technol* 2017;94(9–12):3563–3576.
31. Tao F, Sui F,Liu A,Qi Q, Zhang M, Song B, and Nee AYC. Digital twin-driven product design framework. *Int J Prod Res* 2018;56:1–19.
32. Tuegel EJ, Ingraffea AR, Eason TG, and Spottswood SM. Reengineering aircraft structural life prediction using a digital twin. *Int J Aerospace Eng* 2011;2011:1–14.
33. Uhlemann THJ, Schock C, Lehmann C, Freiberger S, and Steinhilper R. The digital twin: demonstrating the potential of real time data acquisition in production systems. *Proc Manuf* 2017;9:113–120.
34. Singh P, Arora K, Rathore UC, Yang E, Joshi GP, and Son KC. Performance evaluation of grid-connected DFIG-based WECS with battery energy storage system under wind alterations using FOPID controller for RSC. *MDPI J Math* 2023;11(9):2100. https://doi.org/10.3390/math11092100
35. Wang J, Ma Y, Zhang L, Gao RX, and Wu D. Deep learning for smart manufacturing: methods and applications. *J Manuf Syst* 2018;48(48):144–156.
36. Wang X, Ong SK, and Nee AYC. A comprehensive survey of ubiquitous manufacturing research. *Int J Prod Res* 2017a;56(1–2): 604–628.
37. Wang L, Törngren M and Onori M. Current status and advancement of cyber-physical systems in manufacturing. *J Manuf Syst* 2015b;37 (2):517–527.
38. Wu D, Liu S, Zhang L, Terpenny J, Gao RX, Kurfess T, and Guzzo JA. A fog computing-based framework for process monitoring and prognosis in cyber-manufacturing. *J Manuf Syst* 2017;43:25–34.
39. Singh P, Arora K, Rathore UC, Joshi GP, and Cho W. Comparative study of controllers in battery energy storage system integrated with doubly fed induction generator-based wind energy conversion system for power quality improvement. *Energy Rep* 2024;11(9):4587–4600. https://doi.org/10.1016/j.egyr.2024.04.020
40. Zhang C and Sun J. An alternate two phases particle swarm optimization algorithm for flow shop scheduling problem. *Expert Syst Appl* 2009;36(3):5162–5167.

Chapter 5

Evaluation and modelling of the impact of double glass units on the built environment

Amit Kumar Dhir and Pushpendra Kumar Sharma

5.1 INTRODUCTION

Building energy consumption has become a problem, particularly for the envelope, which allows heat to enter the structure. Because of the lack of regulations in emerging nations regarding residential development, contractors continue to accumulate projects in response to the demand for infrastructure. India and other emerging nations have seen tremendous growth in the building sector over the past 10 years. However, residential structures have not been included in the energy study. Over 25% of state-level energy consumption and roughly 23% of worldwide energy consumption are attributed to the residential sector. According to the GBPN report, growth in the energy sector will be nearly eight times greater by 2050 compared to 2012 [1], presenting a challenge as highlighted in the energy audit reports of 2018 [2]. During the COVID-19 pandemic, energy consumption in the residential sector increased due to the shift from offices to homes, as most people began working remotely. For many employees and independent contractors, this has become their primary source of income. Consequently, owners have had to spend most of the previous two years closely monitoring their energy use because of rising power bills for cooling loads. The market for energy conservation measures (ECMs) has become dominant as the trend now moves towards energy efficiency [3]. To compare the best measures against budgeted ones, simulation tools are very helpful and convenient for evaluating different ECMs. When entering data on tenant behaviour, such as window opening and closing habits, caution must be exercised [4]. To some extent, the programme may reflect reality, allowing the tenant to obtain an accurate estimate of the costs associated with specific strategies in order to maximise value for money and payback schedules [5]. Windows play a crucial role in modern living. While large windows are essential for people's well-being and productivity, they also increase energy consumption, complicating supply during peak hours [6]. As a result, power outages are common, making it difficult to operate air conditioners. Therefore, energy conservation and monitoring should be facilitated by IoT-based smart systems [7]. During the design phase, the top roof should

DOI: 10.1201/9781003542568-5

be carefully constructed [8]. A significant remedy to this issue has emerged in recent times with the invention of double glazing windows (DGUs). A thorough examination of the various parameters of DGU [9], such as glass type, orientation, and window-to-wall ratio, is required. Simulation programmes like Design Builder and Energy Plus enable real-site modelling and the study of different ECM types[10], while facilitating analysis and comparisons for recommendations [11]. To understand the causes of discrepancies and ensure that the software is accurately polished and useful for optimal outcomes when implemented in residential buildings, validation of the data from the real site must also be included in the study [12]. The modelling software Design Builder is used in this work, while Energy Plus software is used for simulation. Errors have been analysed, and confirmation is obtained from the electrical records. Low-e glass has been incorporated into the research programme to examine potential outcomes, as it has been a focus of interest for researchers. The product installation was conducted in a residential building selected for the study.

5.2 LITERATURE REVIEW

Both individual well-being and energy consumption are critically influenced by the building envelope. In recent years, due to global warming, countries have become more mindful of their energy usage. The design of the building envelope requires careful consideration of several factors, such as orientation, window size, glazing type, etc. An additional benefit of double glazing units is their ability to reduce the building's energy consumption. Numerous researchers have employed different varieties of low-e glass. Using modelling and the Energy Plus energy simulation programme, it has been determined that double glazing units, hard coated after retrofitting in residential buildings, perform well in tropical environments and can save up to 5.9% of energy use [13]. Numerous ECMs were examined in the building, including replacing the traditional windows with DGUs, which produced encouraging results, and replacing the lighting with LEDs. Simulation software was also used to study the various ECMs, revealing that solar rooftop applications are suitable when the budget exceeds 15 lakhs rupees and the associated costs can be recouped within two years. Research conducted in India found that the budgets for short- and long-term investments were 7 and 25 lakhs rupees, respectively [14].

Approximately 60% of heat losses occur through windows, which should be taken into account when designing them. The author also advocates that the real and simulated values should not differ significantly, as customer feedback indicated a notable discrepancy between the installed product and the performance specified. The simulated and actual values for the DGUs with argon gaps as filler differ by less than 10% [15]. The effectiveness of DGUs and triple glazing windows in reducing heat loss has been studied.

It has been found that while the performance of triple glazing windows in real buildings is similar to that of simulations, there is variability during the summer months, which may be attributed to variations in occupant behaviour and other factors. A test of the sizzling box device was conducted in the lab [16].

Through simulation, it was noted that the building with a two-person household was modelled, and ECMs such as DGU, EPS 100 mm, and EPS 150 mm demonstrated improved outcomes in the building [17]. Using Design Builder software, a simulation for a cold environment was run, and the results showed energy savings of 37%–39% of the cooling loads [18]. After reviewing a long list of current building envelope research projects, the author realised that for architects, engineers, and scientists to unite and support the design of energy-efficient buildings from conceptual to operational stages, they must contribute [19]. Casini provided an overview of the completed articles on dynamic active windows and showcased a variety of previously simulated and tested windows. The author noted that occupant privacy was an issue and that adjustments were needed for the active dynamic windows' visual light transmission (VLT) and switching speed [20]. The author examined the cooling and heating loads in the Indian environment and offered simulated methods to reduce these loads. The exterior wall and roof insulations made of EPS, XPS, and glass wool significantly lower the heating and cooling loads [21]. Yousefi assessed the energy efficiency of several components under investigation and examined the simultaneous effects of glazing factors. It was suggested that argon gas be used to improve insulation and lower cooling loads; in cold areas, payback periods were shown to be shorter than 4.5 years [22]. A previous study [23] investigated the effects of different ECM via simulation and user thermal comfort. The author highlighted the use of air conditioners and discussed their energy consumption. By utilising the proper ECM, electrical energy consumption has been seen to reduce by 12.53%. The application of argon gas-filled double-glazed windows was emphasised in the literature study. One of the problems with DGU is that it cannot be applied to other building types and climates in the same way as it can be in one zone and climate. This requires the use of complex factors for analysis, and ECM should be examined and assessed first. The ECM's cost is significant as it influences the occupant's decision to move towards energy efficiency [24].

5.3 METHODOLOGY

Figure 5.1 illustrates how a residential two-storey structure was modelled in Design Builder using site data in order to determine the building's energy usage. The building envelope is the area of concern since solar ingress was used to quantify losses and gains. As two ECMs, the two varieties of glass have been utilised successively. In the first, standard glass was used as a

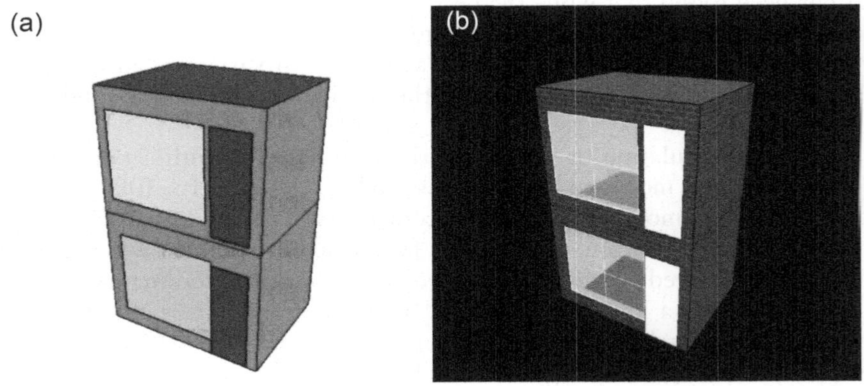

Figure 5.1 (a) Model used in the study. (b) Rendered model.

Table 5.1 Properties of type and their materials for the design builder software

Type	*Material*
Walls	Uninsulated 9-inch walls with plaster 12 mm
Opening	Traditional window, 6 mm clear glass
Flat roof	126 mm RCC
ECM 1	DGU having 6 mm solar silver glazing with argon gas 12 mm
WWR	50%
Window facing	South

comparison to DGU with argon gas, and in the second, air was used as the infill material for DGU. As seen in Figure 5.1, the building's overall size is 1,125 square feet, divided into two storeys that total 2,250 square feet. The study's objectives were to comprehend the building's energy usage. The model was created using the attributes shown in Table 5.1 and the programme Design Builder. It was just a home of two that was occupied. The HVAC schedule for a home setting was from 9 a.m. to 5 p.m. and then from 8 p.m. to 6 a.m. The instance was drawn from a recently established COVID-19 scenario in which the majority of work is done from home. The results concentrated on the building's cooling demand as well as internal solar gain through windows. The warmest day was determined to be June 15, 2021 to provide validation. An Arduino and an LM35 temperature sensor were used to measure the location's temperature. The Energy Plus simulation tool has been used to examine the model's different parameters. Currently, simulation experiments show that the Design Builder and Energy Plus tools are highly useful and closely resemble real-world scenarios. The programme is well-designed to handle scheduling from a residential

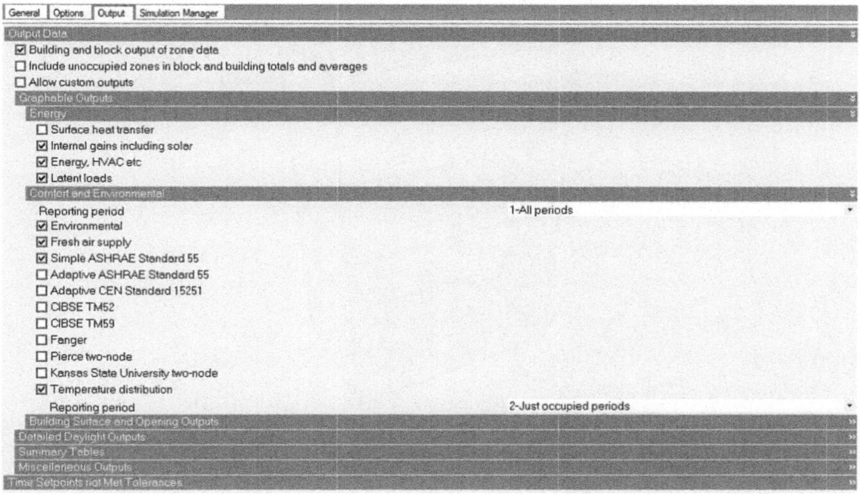

Figure 5.2 Output parameters under study.

standpoint [25,26]. The window's optical characteristics are as follows: for ECM1, they are VLT – 25%, SHGC – 0.3, and a U-value of 1.7; for ECM2, they are VLT – 68%, SHGC – 0.42, and a U-value of 1.9. A 1.5-tonne, three-star worth of separate air conditioners had been seized for research in two of the rooms. Empirical formulae were used to calculate the energy savings for the validation, and 50% occupancy was reported. Figure 5.2's parameters are included in the result.

The simulation was run for a full year to comprehend the advantages and determine the payback period through an economic assessment of the ECM.

5.4 RESULT AND ANALYSIS

The simulation displays the different outcomes of the assessment of the building's cooling load and solar gain entering the envelope through the windows. Comparing the typical window to the higher and lower blocks of the same capacity, both consumption zones have been examined concurrently.

5.4.1 ECM 1

The influence of the DGU with solar glass, which may block heat radiation and thus lower the room's temperature and cooling loads, was demonstrated by ECM 1. Figure 5.3 displays the solar gain. It clearly demonstrates that DGU is the superior window type and has a significant influence on reducing solar gain.

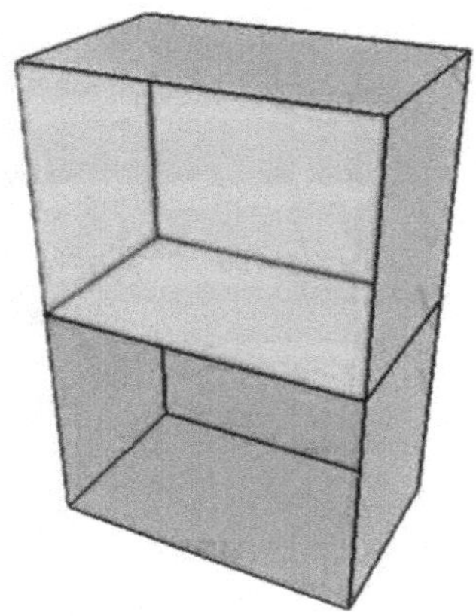

Figure 5.3 Glazing gain on the upper and lower stories at the envelope.

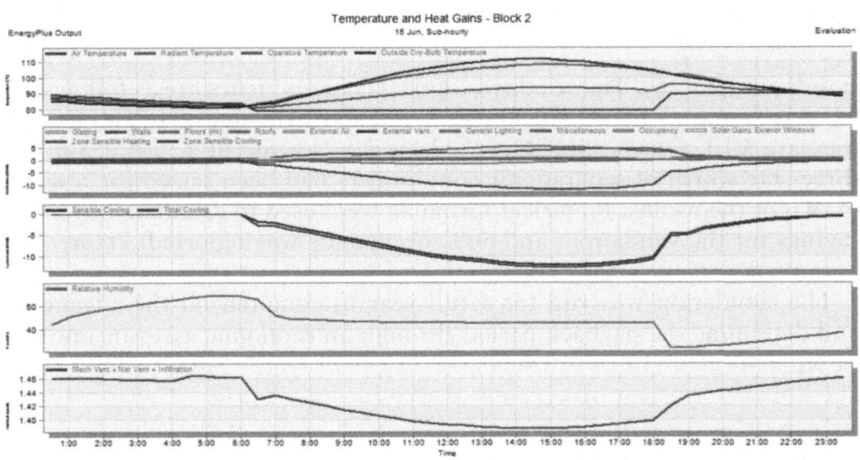

Figure 5.4 Traditional glass window.

Figures 5.4 and 5.5 were used to study the warmest day on the 15 June. The retrofitted ECM 1 installed in the building examined the same warmest day and found that it could lower the inside temperature from the outside to a maximum of 12.9°C.

It was found that the solar gain due to DGU is significantly reduced by 74% for the entire year simulation. The solar gain for ECM 1 DGU was 2561 Kbtu, while the solar gain for conventional glass was 9922 Kbtu.

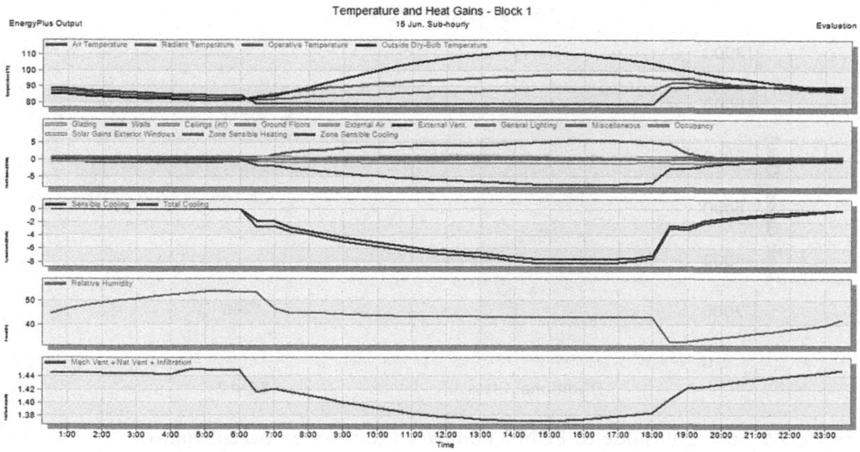

Figure 5.5 ECM 1 as DGU with argon gas.

The simulation used in the study also demonstrated the warmest day. Consequently, a 74% decrease in solar gains was noted.

The sensible cooling for the upper and lower story zones decreased by 36.0% as a result of ECM 1, compared to 19,696 Kbtu for standard glass and 14,473 Kbtu for ECM 1. The analysis demonstrates the significant impact of this decrease on solar gain and cooling. Additionally, the validation revealed a 6.83% error, which might be because of occupant behaviour being well within permissible limits both in the simulation timetable and in real-world experience.

5.4.2 ECM 2

The implementation of ECM 2 in the building with the same model showed the solar gain as indicated in Figure 5.6. ECM 2 has been observed to reduce solar gains by 56%.

Additionally, as Figure 5.7 illustrates, the ECM 2 that uses DGU and air rather than argon gas decreases sensible cooling to 4,932 Kbtu, saving 25.04% on cooling loads.

5.5 ECONOMIC ASSESSMENT

ECM 1 and ECM 2 initially cost 2.75 and 2.19 times, respectively, more than conventional glass. The payback periods for ECM 1 and ECM 2 will be reduced to 1.5 and 1.8 years, respectively, due to the savings in cooling loads.

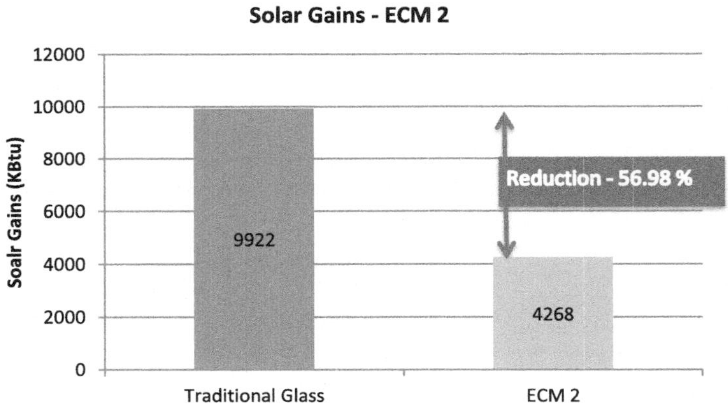

Figure 5.6 Solar gains due to ECM 2 compared to the traditional window.

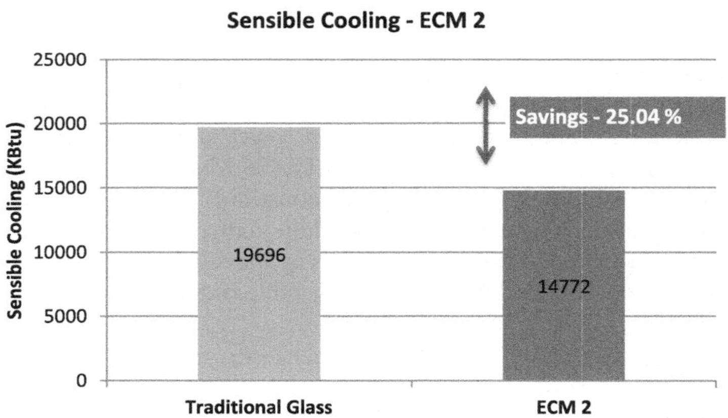

Figure 5.7 Sensible cooling due to ECM 2 compared to the traditional window.

5.6 CONCLUSION

Building loads can be decreased by using an accurate method. The two ECM active solutions for reducing consumption have been the main emphasis of this work. In contrast to conventional glass, the active methods ECM 1 and ECM 2 feature various window formations.

The conclusions drawn were as follows:

ECM 1 reduces cooling loads by 36.0% and lowers solar gains by up to 74%.

ECM 2 reduces cooling loads and solar gains by 25.04% and 56.98%, respectively.

According to the economic evaluation, ECM 1 and ECM 2 have payback periods of 1.5 years and 1.8 years, respectively.

ACKNOWLEDGEMENTS

A special thank you to Mr. Bikramjeet Singh, MD of JKI, a fully automated glass processing operation. In the JKI processing unit, the two window formations are made of Guardian glass. Without the tremendous support, the research project might not have been feasible.

REFERENCES

1. J. Morrissey, "Better data." *Mod. Healthc.*, vol. 27, no. 27, p. 43, 1997.
2. I. – International Energy Agency, "World Energy Outlook EXECUTIVE SUMMARY," 2018, [Online]. Available: www.iea.org/t&c/
3. "Electricity Market Report," *Electr. Mark. Rep.*, no. December, 2020. http://doi.org/10.1787/f0aed4e6-en
4. J. Ferdyn-Grygierek and K. Grygierek, "Optimization of window size design for detached house using trnsys simulations and genetic algorithm," *Archit. Civ. Eng. Environ.*, vol. 10, no. 4, pp. 133–140, 2017. http://doi.org/10.21307/acee-2017-057
5. K. Arora, A. Kumar, V. K. Kamboj, D. Prashar, S. Jha, B. Shrestha, and G. P. Joshi, "Optimization methodologies and testing on standard benchmark functions of load frequency control for interconnected multi area power system in smart grids," *MDPI J. Math.*, vol. 8, no. 6, p. 980, 2020. http://doi.org/10.3390/math8060980
6. K. H. Cheong, Y. H. Teo, J. M. Koh, U. R. Acharya, and S. C. Man Yu, "A simulation-aided approach in improving thermal-visual comfort and power efficiency in buildings," *J. Build. Eng.*, vol. 27, no. August 2019, p. 100936, 2020. http://doi.org/10.1016/j.jobe.2019.100936
7. A. Kumar, S. Sharma, N. Goyal, A. Singh, X. Cheng, and P. Singh, "Secure and energy-efficient smart building architecture with emerging technology IoT," *Comput. Commun.*, vol. 176, no. May, pp. 207–217, 2021. http://doi.org/10.1016/j.comcom.2021.06.003
8. M. S. Jaral, and S. V. Singh, "Recent advances in sustainable technologies," in *Recent Advances in Sustainable Technologies*, Singapore: Springer, 2021, pp. 165–181.
9. A. K. Dhir, "Role of suitable glazing in minimizing energy consumption in Jalandhar region," *Int. J. Res. Anal.*, vol. 5, no. 4, pp. 1527–1530, 2018.
10. B. Bharath, A. Sami, and K. Rastogi, "Analysis of a structure for efficient energy utilization using design builder," *IOP Conf. Ser. Mater. Sci. Eng.*, vol. 1017, no. 1, 2021. http://doi.org/10.1088/1757-899X/1017/1/012016
11. M. Alhashmi, G. Chhipi-Shrestha, R. Ruparathna, K. M. Nahiduzzaman, K. Hewage, and R. Sadiq, "Energy performance assessment framework for residential buildings in Saudi Arabia," *Sustain.*, vol. 13, no. 4, pp. 1–25, 2021. http://doi.org/10.3390/su13042232
12. R. Andersen, V. Fabi, J. Toftum, S. P. Corgnati, and B. W. Olesen, "Window opening behaviour modelled from measurements in Danish dwellings," *Build. Environ.*, vol. 69, pp. 101–113, 2013. http://doi.org/10.1016/j.buildenv.2013.07.005

13. S. Somasundaram, S. R. Thangavelu, and A. Chong, "Effect of existing façade's construction and orientation on the performance of low-E-based retrofit double glazing in tropical climate," *Energies*, vol. 13, no. 8, 2020. http://doi.org/10.3390/en13082016
14. S. L. Swati Dixit, "Energy Conservation using Window Glazing in India," *Int. J. Adv. Res. Electr. Electron. Instrum. Eng.*, vol. 4, no. 11, pp. 8645–8654, 2015. http://doi.org/10.15662/ijareeie.2015.0411016
15. E. Cuce, "Accurate and reliable U-value assessment of argon-filled double glazed windows: A numerical and experimental investigation," *Energy Build.*, vol. 171, pp. 100–106, 2018. http://doi.org/10.1016/j.enbuild.2018.04.036
16. A. Rana et al., "(11) (PDF) Thermal performance of double and triple glazed windows Experimental results from lab and in-situ measurements," no. 1999, pp. 1–7, 2018, [Online]. Available: https://www.researchgate.net/publication/326316386_Thermal_performance_of_double_and_triple_glazed_windows_Experimental_results_from_lab_and_in-situ_measurements
17. Ü. Bahadır, X. Thomollari, and V. Toğan, "Evaluation of energy-cost efficient design alternatives for residential buildings," *J. Constr. Eng. Manag. Innov.*, vol. 1, no. 1, pp. 43–54, 2018. http://doi.org/10.31462/jcemi.2018.01043054
18. Y. Yaşar and S. M. Kalfa, "The effects of window alternatives on energy efficiency and building economy in high-rise residential buildings in moderate to humid climates," *Energy Convers. Manag.*, vol. 64, no. 4, pp. 170–181, 2012. http://doi.org/10.1016/j.enconman.2012.05.023
19. Y. Luo, L. Zhang, M. Bozlar, Z. Liu, H. Guo, and F. Meggers, "Active building envelope systems toward renewable and sustainable energy," *Renew. Sustain. Energy Rev.*, vol. 104, no. July 2018, pp. 470–491, 2019. http://doi.org/10.1016/j.rser.2019.01.005
20. M. Casini, "Active dynamic windows for buildings: A review," *Renewable Energy*, vol. 119, pp. 923–934, Apr. 01, 2018. http://doi.org/10.1016/j.renene.2017.12.049
21. K. Arora, A. Kumar, V. K. Kamboj, D. Prashar, B. Shrestha, and G. P. Joshi, "Impact of renewable energy sources into multi area multi-source load frequency control of interrelated power system," *MDPI J. Math.*, 9(2), p. 186, 2021. https://doi.org/10.3390/math9020186
22. F. Yousefi, Y. Gholipour, Y. Saboohi, and W. Yan, "Interaction of glazing parameters, climatic condition and interior shadings: performing energy and cost analysis in a residential building in Iran," *Energy Effic.*, vol. 13, no. 1, pp. 159–176, 2020. http://doi.org/10.1007/s12053-019-09831-w
23. A. Esertine and Siddhartha, "Energy conservation strategies in residential building-A case study," *AIP Conf. Proc.*, vol. 2273, November, 2020. http://doi.org/10.1063/5.0024301
24. K. Arora, G. P. Joshi, M. Ragab, M. Rawa, A. H. Milyani, R. F. Mansour, and E. Yang, "Bilateral contract for load frequency and renewable energy sources using advanced controller", *Comp., Mat. Continua*, vol. 73, no. 2, pp. 3165–3180, 2022. http://doi.org/10.32604/cmc.2022.026966

25. G. Kirankumar, S. Saboor, and T. P. Ashok Babu, "Thermal analysis of wall and window glass materials for cooling load reduction in green energy building design," *Mater. Today Proc.*, vol. 4, no. 9, pp. 9514–9518, 2017. http://doi.org/10.1016/j.matpr.2017.06.215
26. S. Shaik, K. Gorantla, A. Ghosh, C. Arumugam, and V. R. Maduru, "Energy savings and carbon emission mitigation prospective of building's glazing variety, window-to-wall ratio and wall thickness," *Energies*, vol. 14, no. 23, p. 8020, 2021. http://doi.org/10.3390/en14238020

Chapter 6

Estimation for establishing a sustainable solar-powered civilization

Jumi Bharali and Krishan Arora

6.1 INTRODUCTION

The most common energy source in use today is electrical energy. We consume enough energy each day to meet our fundamental demands if we take a closer look at the things we encounter in our daily lives. However, why do we typically use electricity? It's actually quite easy to understand! One of the most efficient forms of energy is electrical energy. When compared to all other forms of energy, the production-to-utilization ratio is extremely high.

According to data on electrical energy use, we utilize approximately 17.4TW, or 17400,00,00,00,000 W, of energy every day. That astounding figure demonstrates how much energy we consume. Power consumption in India witnessed a year-on-year rise of nearly 8%, reaching 1,221.15 billion units during the April–December period 2023–2024. This underscores the extent of the nation's electrical energy needs. Currently, conventional sources of energy account for the majority of India's electrical energy usage, while the supply of non-conventional energy sources remains limited. The purpose of this project is to demonstrate how we can rely solely on non-conventional energy sources to power our nation.

6.2 CONCEPT OF SOLAR SOCIETY

Solar Society is an imaginary community that we created, entirely powered by solar energy. This society is self-sufficient in terms of power generation and utilization and does not depend on any external sources for electrical power. Figure 6.1 shows the 2D layout of the society plan.

The society is equipped with all the basic facilities. It includes a marketplace, an auditorium hall, water supply system, playground, central park, and mini-gardens. All of the power needed for these facilities is supplied by solar energy. This society aims to be a completely zero-carbon emission society in terms of electrical power.

 DOI: 10.1201/9781003542568-6

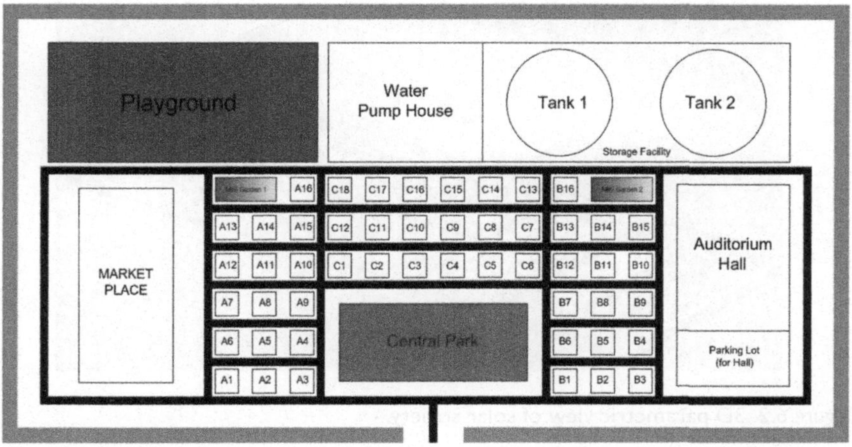

Figure 6.1 Society layout.

The main parts of the society are given as follows:

1. 50 houses
2. A market place with 10 shops
3. An auditorium hall with a parking lot
4. Water supply system
5. Street lighting
6. Central park, playground, mini-garden

Figure 6.2 shows a 3D parametric view of the society layout and its installation. The society's main facilities can be clearly seen in the image, giving us a glimpse of what the society will actually look like in real life.

Figure 6.3 provides a drone perspective of the society, allowing us to see the society and its facilities from the front.

6.3 HOUSES IN SOLAR SOCIETY

There are 50 houses in the solar society. Each house is identical and has the same installations. Each house includes a living room, 2 rooms, a kitchen, a bathroom, a veranda, and a power house for solar equipment installation (Figures 6.4 and 6.5).

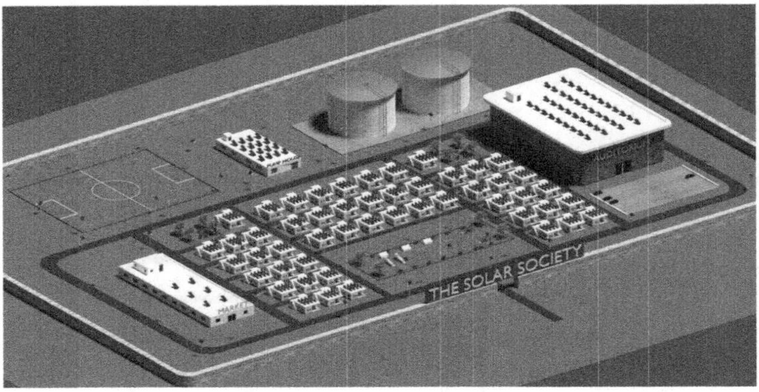

Figure 6.2 3D parametric view of solar society.

Figure 6.3 Drone perspective view of solar society.

6.3.1 3D Parametric view of house

Figure 6.4 3D parametric view of house.

6.3.2 Plan diagram of house

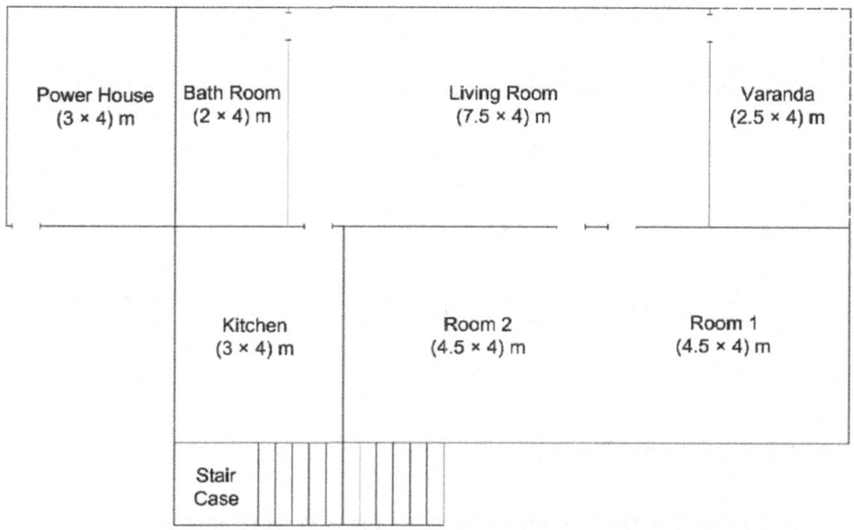

Figure 6.5 Plan diagram of house.

6.4 MARKETPLACE OF SOLAR SOCIETY

The marketplace is a large complex with 10 shops. All the shops are equipped with identical electrical installations. It features a corridor in the middle for buyers to move around the complex. The terrace portion has the solar panels and the powerhouse (Figures 6.6 and 6.7).

6.4.1 3D parametric view of marketplace

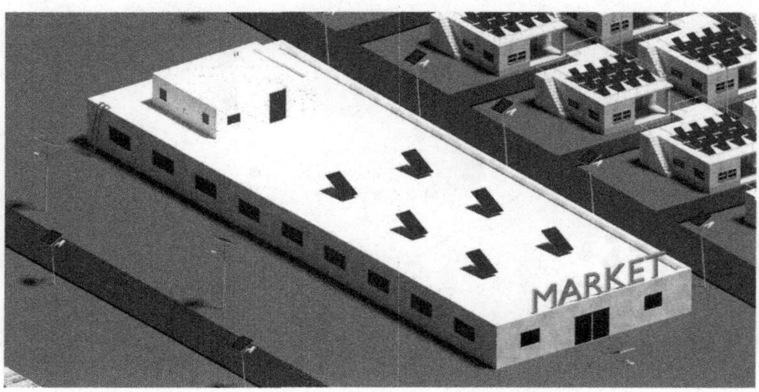

Figure 6.6 3D parametric view of marketplace.

6.4.2 Plan diagram of marketplace

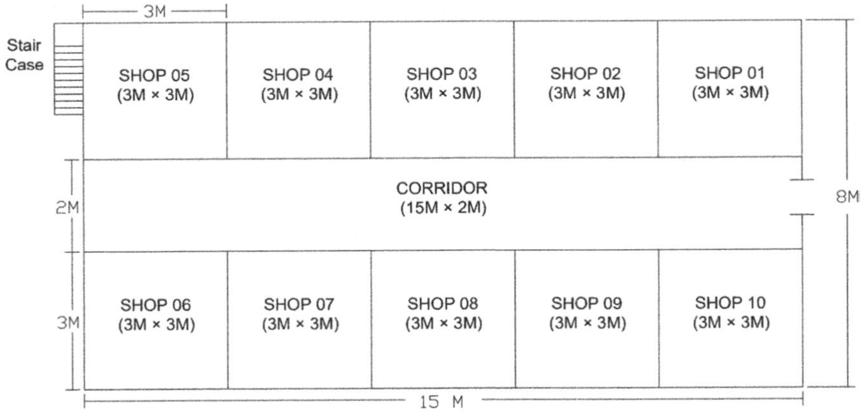

Figure 6.7 Plan diagram of marketplace.

6.5 WATER SUPPLY SYSTEM OF SOLAR SOCIETY

The water supply system of the solar society mainly consists of a pumphouse and two storage tanks. The pumphouse is equipped with five 2 HP motors for pumping water from the ground to storage facilities (Figures 6.8 and 6.9).

6.5.1 3D parametric view of water supply system

Figure 6.8 3D parametric view of water supply system.

6.5.2 Plan diagram of pump house

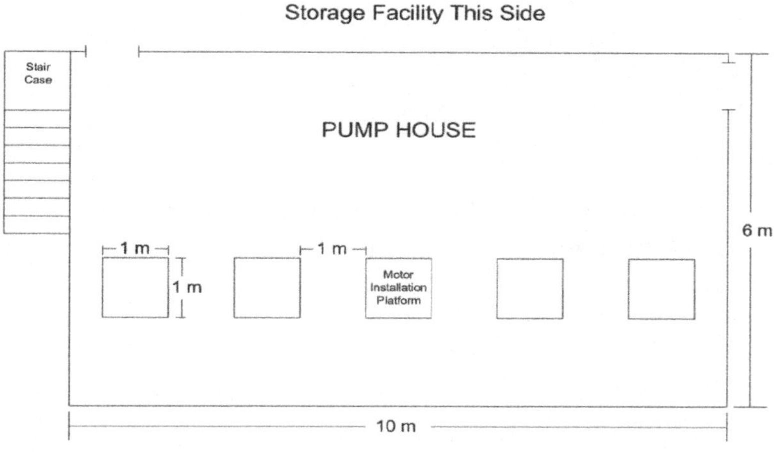

Figure 6.9 Plan diagram of pump house.

6.6 AUDITORIUM OF SOLAR SOCIETY

The auditorium of the solar society is a large public gathering hall with a total floor area of (15×12) m. The seating space is about (12×10) m or 120 m^2. It has a stage with dimensions of (5×6) m, as well as one green room and a storeroom. The terrace portion is installed with solar panels and a powerhouse (Figures 6.10 and 6.11).

6.6.1 3D parametric view of auditorium

Figure 6.10 3D parametric view of auditorium.

6.6.2 Plan diagram of auditorium

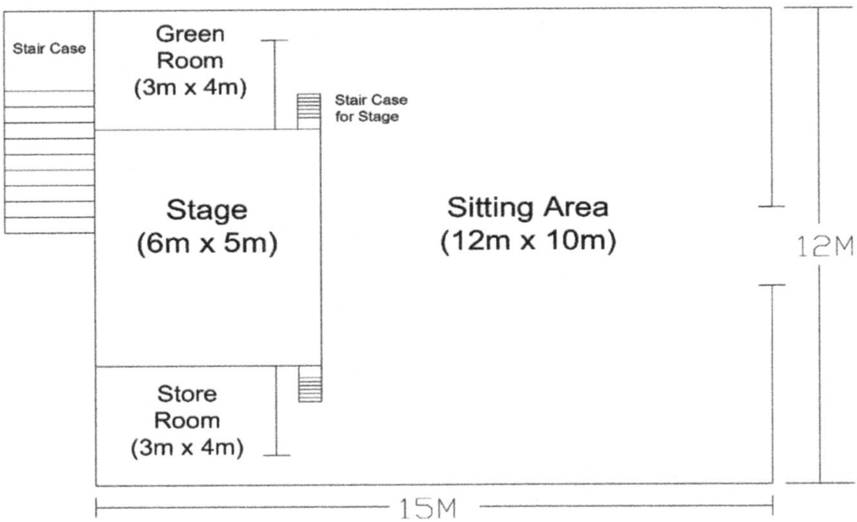

Figure 6.11 Plan diagram of auditorium.

6.7 ESTIMATING PROCEDURE

The estimate for the whole society has been divided into five main parts, which are as follows:

1. House
2. Market place
3. Pump house
4. Auditorium
5. Street lighting

To make the estimate, we first created the plan diagram for the installation, calculated the total load connected in that sector, and determined the number of circuits that would be required. Then, using the plan diagram, we calculated the amount of different types of materials required. In the final stage, we calculated the total cost required for the installation of equipment and accessories.

6.8 ESTIMATION FOR HOUSE

6.8.1 Installation plan diagram (Figure 6.12)

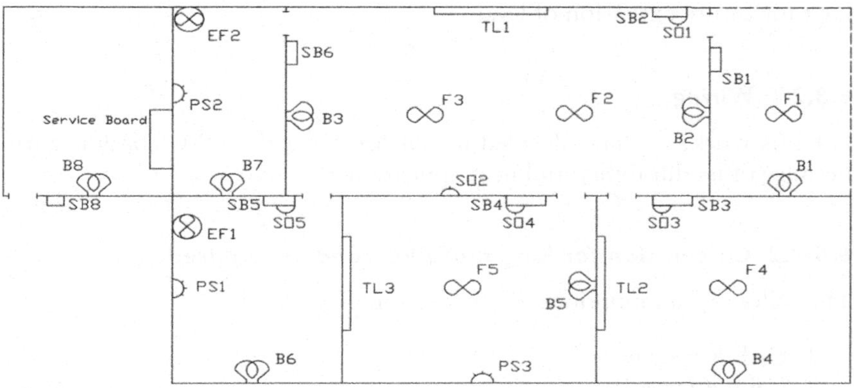

Figure 6.12 House installation plan *(with reference to Figure 5.5).*

6.8.2 Load calculation

Total Lighting Load: (140+480+500) W=1,120 W=approx. 1,200 W

Total Power Load: 4,000 W

Now, since the total lighting load exceeds 800 W, we will require two sub-circuits to supply the lighting load, each with a total load of 700 and 500 W respectively. Additionally, since the power load exceeds 2,000 W, we will require two circuits to supply the power load, each with a load of 2,000 W (Table 6.1).

Table 6.1 Load calculation for house

Location	*Light (W)*	*Fan (W)*	*Socket (W)*	*Power socket*
Veranda	1×10	1×80		
Room 1	1×10 1×20	1×80	1×100	
Room 2	1×10 1×20	1×80	1×100	1×1,000
Kitchen	1×10	EF: 1×40	1×100	1×1,000
Bathroom	1×10	EF: 1×40		1×2,000
Living room	2×10 1×20	2×80	2×100	
Power house	1×10			
Total	**140**	**480**	**500**	**4,000**

Therefore, the total number of circuits required will be four. According to IE rules, the permissible load in the lighting circuit of a house is 800 W, so the installed load capacity of the house will be (800+800+2,000+2,000=5,600) W. However, the connected load is around 5,200 W. The rest is kept for future extension of load.

6.8.2.1 Wiring

For this work, we have decided to opt for concealed PVC conduit wiring because of its durability and neat appearance.

6.8.2.2 Calculation for length of PVC conduit required

The following assumptions are made from floor level:

1. Ceiling=3.5 m
2. Ceiling fan= 3 m
3. Light points= 3 m
4. D.B.=1.8 m
5. M.S.=1.8 m
6. S.B.=1.5 m
7. Light plug=1.5 m

Therefore, the total length of conduit required is as follows (Table 6.2):

Table 6.2 Length of conduit for house

Location	*Horizontal run (in m)*	*Vertical drop to switch (in m)*	*Vertical rise to ceiling fan (in m)*	*Vertical rise to bulb (in m)*	*Total (in m)*
Veranda	3+1.5	1	1	0.5	7
Living room	7.5+7.5+2	1	2	1	21
Room 1	3+2.5+4	1	1	0.5	12
Room 2	3+2+2	1	1	0.5	9.5
Kitchen	3+2+2	1	1	0.5	9.5
Bathroom	4+2+1	1	1	0.5	9.5
Power house	1.5+1.5	1	1	0.5	5.5
Total					74
For wall crossing (5%)					3.7≈4
Wastage (10%)					7.4≈8
Grand total					86 m

6.8.2.3 Calculation for size and length of wire from distribution box to switchboards

Here, we will be using PVC insulated copper wire of the appropriate size according to load.

For size of wire, refer to Table 6.3.

For length of wire, refer to Table 6.4.

6.8.2.4 Calculation for length of wire from switchboard to appliances

Here also, we will be using PVC insulated copper wire for connections. Keeping in mind the load, we will use 1 and 2 mm^2 wires, as necessary. For all purposes, we will use a tee or joint box system of wiring.

For the length of wire, refer to Table 6.5.

Table 6.3 Size of wire from distribution box to switchboard of house

Circuit	*Load (P) (in watts)*	*Supply voltage (V) (in volts)*	*Load current (P/V) (in amps)*	*Size of wire (in mm2)*
Circuit 1	800	230	3.5	2.5
Circuit 2	800	230	3.5	2.5
Circuit 3	2,000	230	8.70	3.5
Circuit 4	2,000	230	8.70	3.5

Table 6.4 Length of wire from distribution box to switchboard of house

Circuit	*Switch board*	*Length of wire (in m)*	*Total length (in m)*
Circuit 1	SB1	14	34×3=102
	SB3	11	
	SB4	9	
Circuit 2	SB 2	11	18×3=54
	SB 5	3	
	SB 6	4	
Circuit 3	PS 2	1	11×3=33
	PS 3	10	
Circuit 4	PS 1	3	3×3=9
Total length of 2.5 mm^2 conductor			156
Total length of 3.5 mm^2 conductor			42

Table 6.5 Calculation for length of wire from switchboard to appliance

Room (SB)	*Appliance*	*Size of wire (in mm2)*	*Distance form SB to appliance (d) (in m)*	*Length of wire (d × 2) (in m)*
Veranda (SB1)	B1	1	2.5	12
	F1		3.5	
Living room (SB2)	B2	1	12	115
	B3		12	
	TL1		4.5	
	F2		9	
	F3		9	
	SO2		11	
Room 1 (SB3)	B4	1	2.5	28
	F4		5	
	TL2		6.5	
Room 2 (SB4)	B5	1	2.5	28
	F5		5	
	TL3		6.5	
Kitchen (SB5)	EF1	1	4.5	18
	B6		4.5	
Bathroom (SB6)	EF2	1	4	15
	B7		3.5	
Power house (SB7)	B8	1	3.5	7
Total				223

Hence,

$$\text{Total length of conductor} = 223\,\text{m}$$

$$\text{Considering extra } 10\% \text{ for wastage and wall crossing} = 22.3\,\text{m}$$

Therefore,

$$\text{Total length of wire required will be} = 246.3\,\text{m}$$

$$= 250\,\text{m}(\text{approx})$$

6.8.3 Solar installation planning for the house

In the house, the total installed load capacity is around 5,600 W. Therefore, the design of the solar installation must be done to meet the energy needs of 5,600 W. All the solar equipment must fulfill the criteria of serving the total load.

6.8.3.1 Solar panels

The solar panel that we will be using is Luminous's mono-crystalline 380 W-24 V panel. To supply the load, we will need 15 such panels, which will serve a total load of 5,700 W.

6.8.3.2 Solar inverter

Since our total installed load capacity is about 5,600 W (5.6 kW), we will be using Flin Energy's hybrid inverter with a rating of 7 kW – 24 V/240 V, keeping provision for future load extension. This inverter is best suited for our installation.

6.8.3.3 Solar charge controller

The total load current of our house is (5,600/230=24.5 A), approximately 25 A. Therefore, we will be using Luminous's MPPT-type solar charge controller with a 30 A rating, allowing for future load extension.

6.8.3.4 Battery bank

As the total load current is about 25 A, considering a battery run time of 48 hours with full load current, we will need a battery bank of 1,200 Ah.

Therefore, to create this battery bank, we will use 8 luminous lead acid batteries, each rated at 150 Ah-24 V, all connected in a parallel configuration.

6.8.3.5 Grounding/earthing system

For grounding purposes, we will be using a pipe earthing system. In this type of earthing, a galvanized steel perforated pipe is buried vertically, connecting all the electrical conductors to the earth, where the depth of the pipe depends on the soil conditions. Pipe earthing is a more economical type of earthing compared to other methods. Earth resistance must be less than or equal to 1 Ω (Tables 6.6 and 6.7).

6.8.3.6 Cost of materials

Table 6.6 Cost of materials for house

Particulars	*Quantity*	*Cost/Unit*	*Total cost (in ₹₹)*
Wiring and accessories			
PVC conduit (50 mm dia.)	86 m	45/m	3,870
PVC conduit T and L bends	50 Nos.	5/pc	250
3.5 mm² PVC Cu wire	42 m	56/m	2,352
2.5 mm² PVC Cu wire	156 m	45/m	7,020

(*Continued*)

Table 6.6 (Continued) Cost of materials for house

Particulars	*Quantity*	*Cost/Unit*	*Total cost (in ₹)*
1 mm^2 PVC Cu wire	250 m	20/m	5,000
PVC junction box	8 Nos	50/pc	400
PVC switchboard	7 Nos	100/pc	700
MCB	4 Nos	250/pc	1,000
Four-way DB	1 Nos	500	500
Switch	26 Nos	55/pc	1,430
Socket	7 Nos	80/pc	560
Ceiling roses	10 Nos	35/pc	350
Batten holder	7 Nos	50/pc	350
Ceiling fans	5 Nos	2,000/unit	10,000
Exhaust fan	2 Nos	1,500/unit	3,000
Tube light	3 Nos	600/unit	1,800
Bulb	8 Nos	200/unit	1,600
Fan regulator	5 Nos	350/unit	1,750
GI earth pipe	1 Nos	1,000/unit	1,000
Earthing materials		1,500	1,500
Solar installation			
Solar panels	15 Nos.	15,500/unit	232,500
Solar inverter	1 Nos.	97,000/unit	97,000
Solar charge controller	1 Nos.	950/unit	950
Battery	8 Nos.	13,500/unit	1,08,000
Panel wires	30 m	15/m	450
Solar panel mounting	15 Nos.	2,500/unit	37,500
Total			520,832 ≈ 520,850

Table 6.7 Total cost of installation for house

Material cost	₹ 520.850
Labor charges (10% of material cost)	₹ 52.085
Super vision charges (2% of material cost)	₹ 10.417
Contingency charges (5% of material cost)	₹ 26.042
Grand total	₹ 609,394

6.8.3.7 Total cost of installation

The total cost of electrical installation in one house is approximately ₹ 6,10,000. Therefore, the total installation cost for all 50 houses will be ₹ 3,05,00,000

6.8.4 Estimation for marketplace

6.8.4.1 Installation plan diagram (Figure 6.13)

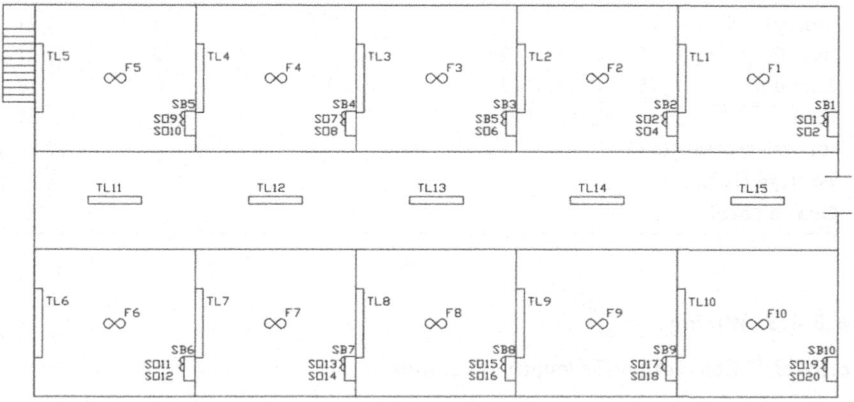

Figure 6.13 Market installation plan.

6.8.4.2 Load calculation

The total load for all the appliances in the marketplace is 3,100 W. Considering all the load factors, we will be using five circuits and; therefore, a five-way distribution box. The five circuits are divided as follows (Table 6.8):

Circuit 1: Shop 01 and Shop 02, Total Load = 600 W
Circuit 2: Shop 03 and Shop 04, Total Load = 600 W
Circuit 3: Shop 05 and Shop 06, Total Load = 600 W
Circuit 4: Shop 07 and Shop 08, Total Load = 600 W
Circuit 5: Shop 09, Shop 10 and Corridor, Total Load = 700 W

As the permissible load for each lighting circuit is 800 W, the total installed capacity for five circuits is 4,000 W. However, the total connected equipment load is only around 3,100 W, allowing for potential future load extensions.

Table 6.8 Load calculation of marketplace

Location	*Tube light*	*Fan*	*Socket*	*Total*
Shops (01–05)	5×1×20 W	5×1×80 W	5×2×100 W	1,500 W
Shops (06–10)	5×1×20 W	5×1×80 W	5×2×100 W	1,500 W
Corridor	1×5×20 W	–	–	100 W
Total	300 W	800 W	2,000 W	3,100 W

Table 6.9 Length of conduit for marketplace

Location	*Horizontal run (in m)*	*Vertical drop to switch (in m)*	*Vertical rise to ceiling fan (in m)*	*Vertical rise to tube light (in m)*	*Total (in m)*
Shop 01–05	15	2	2	2	21
Shop 06–10	15	2	2	2	21
Corridor	15	2	0	0	17
Total					59
For wall crossing (5%)					2.9≈3
Wastage (10%)					5.9≈6
Grand total					68 m

6.8.4.3 Wiring

6.8.4.3.1 Calculation for length of conduit

The following assumptions are for the floor:

1. Ceiling=3.5 m
2. Ceiling fan=3 m
3. Tube Light=3 m
4. Switchboard=1.5 m

Therefore, length of conduit required will be as follows (Table 6.9):

6.8.4.3.2 Calculation for size of wire

In the marketplace, the maximum load on a single circuit is 800 W. Calculating the load current, we find that the maximum load current of a single circuit is approximately 3.5 Amps, or roughly 4 Amps. To maintain uniformity in wire size throughout the marketplace, we will use 2.5 mm^2 PVC insulated copper wire.

6.8.4.3.3 Calculation for length of wire from distribution box to switchboard

Calculation for length of wire from switchboard to appliance (Tables 6.10 and 6.11):

6.8.5 Solar installation planning of the marketplace

In the marketplace, the total installed load capacity is around 4,000 W. So, the design of the solar installation must be done to meet the need of 4,000 W for energy. All the solar equipment must fulfill the criteria of serving the total load.

Table 6.10 Calculation for length of wire from distribution box to switchboard for marketplace

Circuit	*Switch board*	*Distance (in m)*	*Length of wire (in m)*
Circuit 1	SB1 and SB2	19	19×3=57
Circuit 2	SB3 and SB4	13	13×3=39
Circuit 3	SB5 and SB6	10	10×3=30
Circuit 4	SB7 and SB8	13	13×3=39
Circuit 5	SB9, SB10, SB11	19+17	19×3+17×2=91
Total			256
15% Extra for wall crossing and wastage			38.4≈39
Grand total			295

Table 6.11 Calculation for length of wire from switchboard to appliances for marketplace

Location	*Appliance*	*Distance (in m)*	*Length of wire (in m)*	*Total length*
Shops	Tube light	6	6×2=12 (for 1 shop)	120
	Fan	3.5	3.5×2=7 (for 1 shop)	70
Corridor	Tube light	15	15×2=30	30
Total				220
15% Extra for wall crossing and wastage				33
Grand total				253≈255

6.8.5.1 Solar panels

The solar panel that we will be using is the Luminous's mono-crystalline 380 W-24 V Panel. To supply the load, we will need 11 such panels, which will serve a total load of 4,180 W.

6.8.5.2 Solar inverter

Since our total installed load capacity is about 4,000 W (4 kW), we will be using Flin energy's hybrid inverter with a rating of 5 kW – 24 V/240 V, allowing for future load extension.

6.8.5.3 Solar charge controller

The total load current of our house is approximately (4,000/230=17.3 A), which we will round up to 20 A. Therefore, we will use Luminous's MPPT type solar charge controller with a 30 A rating, providing for future load extension.

6.8.5.4 Battery bank

Since the total load current is about 20 A and considering a battery runtime of 36 hours with full load current, we will need a battery bank of 720 Ah.

To make up this battery bank, we will use five luminous lead acid batteries with a rating of 150 Ah-24 V, all connected in parallel.

6.8.5.5 Grounding/earthing system

Here, in the marketplace, we will implement pipe earthing because it is economical and durable (Table 6.12).

6.8.5.6 Cost of materials

Table 6.12 Cost of materials for marketplace

Particulars	*Quantity*	*Cost/Unit*	*Total cost (in ₹)*
Wiring and accessories			
PVC conduit (50 mm Dia.)	68 m	45/m	3,060
PVC conduit T and L bends	35 Nos.	5/pc	175
2.5 mm² PVC Cu wire	550 m	45/m	24,750
PVC junction box	11 Nos	50/pc	550
PVC switchboard	11 Nos	100/pc	1,100
MCB	5 Nos	250/pc	1,250
5-Way DB	1 Nos	500/pc	500
Switch	45 Nos	55/pc	2,475
Socket	20 Nos	80/pc	1,600
Ceiling roses	15 Nos	35/pc	525
Ceiling fans	10 Nos	2,000/unit	20,000
Tube light	15 Nos	600/unit	9,000
Fan regulator	10 Nos	350/unit	3,500
GI earth pipe	1 Nos	1,000/unit	1,000
GI earthing strip	10 m	35/m	350
Earthing materials	—	1,500	1,500
Solar installation			
Solar panels	11 Nos.	15,500/unit	170,500
Solar inverter	1 Nos.	75,000/unit	75,000
Solar charge controller	1 Nos.	950/unit	950
Battery	5 Nos.	13,500/unit	67,500
Panel wires	20 m	15/m	300
Solar panel mounting	11 Nos.	2,500/unit	27,500
Total			413,085 ≈ 414,000

6.8.5.7 Total cost of installation

Table 6.13 Total cost of installation for marketplace

Material cost	₹ 414,000
Labor charges (10% of material cost)	₹ 41,400
Super vision charges (2% of material cost)	₹ 8,280
Contingency charges (5% of material cost)	₹ 20,700
Grand total	₹ 484,380

Therefore, The total cost of electrical installation for the marketplace is around ₹4,85,000 (Table 6.13).

6.8.6 Estimation for pump house

6.8.6.1 Installation plan diagram (Figures 6.14 and 6.15)

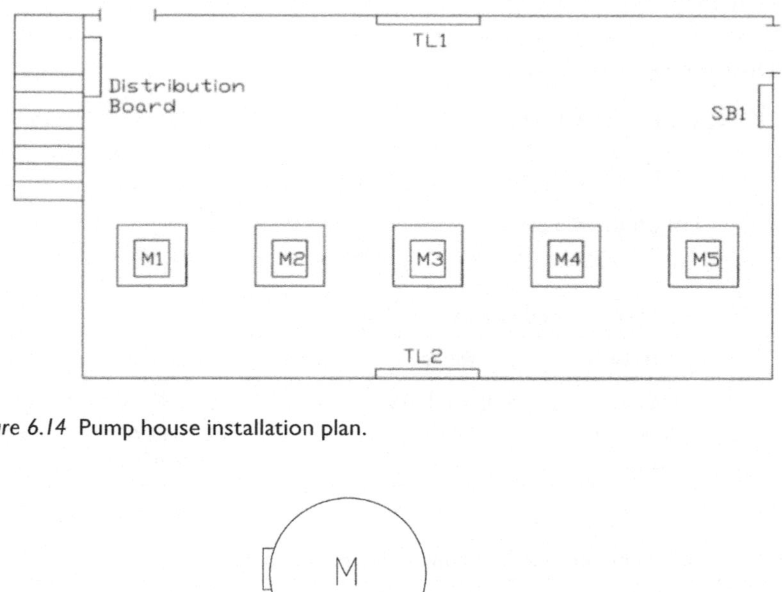

Figure 6.14 Pump house installation plan.

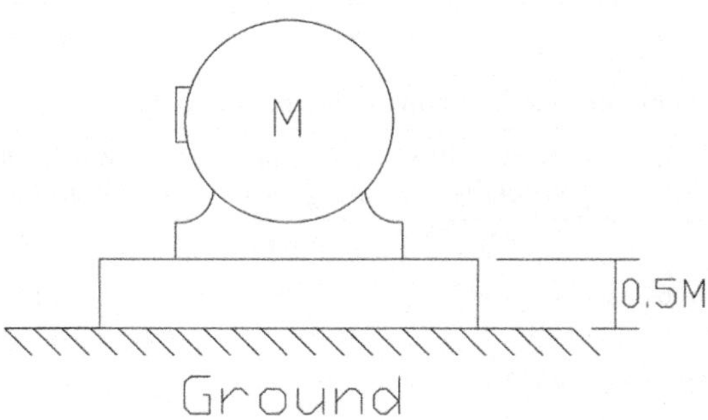

Figure 6.15 Motor installation plan.

6.8.6.2 Load calculation

Here, the total load is around 7,500 W. We will divide the load into six main circuits: five circuits for each motor (where the MCB will itself work as a switch) and one circuit for the tube lights. Therefore, we will be using a six-way distribution box (Table 6.14).

6.8.6.3 Wiring

We have decided to use PVC Concealed Conduit type wiring. Since this area primarily deals with water, there is a risk of water causing a short circuit in the house. Therefore, the wiring needs to be thoroughly checked for any non-insulated parts.

6.8.6.3.1 Calculation for length of conduit required

The following assumptions are made from the floor: Ceiling=3.5 m (Table 6.15)

$$\text{Distribution box} = 2\,\text{m}$$

$$\text{Switchboard} = 1.5\,\text{m}$$

$$\text{Motor} = 0.5\,\text{m}$$

$$\text{Tube light} = 3\,\text{m}$$

Table 6.14 Load calculation of pump house

Particulars	*Wattage*	*Quantity*	*Total load*
Motor	2 HP=1,490 W	5	7,450 W
Tube lights	20 W	2	40 W
Total			7,490 W

Table 6.15 Calculation for length of conduit for pump house

Particulars	*Horizontal run (in m)*	*Vertical drop to floor (in m)*	*Vertical rise to ceiling/motor (in m)*	*Vertical drop to t. light (in m)*	*Total (in m)*
Motor	10	2	0.5×5	–	14.5
Tube light	6+6	–	1+2	0.5×6	17
	Total				31.5≈32
	For wall crossing (5%)				1.6≈2
	Wastage (10%)				3.2≈4
	Grand total				38≈40 m

6.8.6.3.2 Calculation for size of conductor

As mentioned earlier, the circuits are primarily divided into two main categories: motors and tube lights, for which we will use different sizes of conductors.

For the motors, the wattage of each motor is approximately 1,490 W. The full load current in the running stage of the motor is approximately (1,490/230 =6.25 Amps) 6.5 Amps. However, the starting current is always higher in a motor, we will consider the current to be 13 Amps. For this reason, we will use a 5 mm^2 PVC insulated copper conductor.

For the tube lights, the total load is only 40 W, resulting in a full load current of approximately 0.5 Amps. Therefore, we will use 1 mm^2 PVC insulated copper conductor (Table 6.16).

6.8.6.3.3 Calculation for length of conductor

Table 6.16 Calculation for length of conductor for pump house

Appliance		*Size of conductor (in mm²)*	*Length of conductor (in m)*	*Total length (in m)*
From	*To*			
DB	M1	5	3×3	9
	M2		5×3	15
	M3		7×3	21
	M4		9×3	27
	M5		11×3	33
DB	SB1	1	(10+3)×2	26
SB1	TL1	1	12×2	24
	TL2		12×2	24
Total length of 2.5 mm² conductor				105
Total length of 2.5 mm² conductor with 15% for wastage				120.75 ≈ 120
Total length of 1 mm² conductor				74
Total length of 1 mm² conductor with 15% for wastage				85.1 ≈ 85

6.8.7 Solar installation planning for the pump house

In the pump house, the total load is around 7,500 W. Therefore, the design of the solar installation must be done to meet the need for 7,500 W of energy. All the solar equipment must fulfill the criteria of serving the total load.

6.8.7.1 Solar panels

The solar panel we will be using is Luminous's mono-crystalline 380 W-24 V panel. To supply the load, we will need 11 such panels, which will serve a total load of 7,600 W.

6.8.7.2 *Solar inverter*

Since our total load is about 7,500 W (7.5 kW), we will be using Flin Energy's Hybrid Inverter with a rating of 8 kW–24 V/240 V, allowing for future load extension.

6.8.7.3 *Solar charge controller*

The total load current of our house is approximately (7,500/230=32.6 A), which we will round up to 20 A. Therefore, we will use Luminous's MPPT-type solar charge controller with a 30 A rating, providing for future load extension.

6.8.7.4 *Battery bank*

Since the total load current is about 20 A and considering a battery runtime of 36 hours with full load current, we will need a battery bank of 720 Ah.

To make up this battery bank, we will use 5 luminous lead acid batteries with a rating of 150 Ah-24 V, all connected in parallel.

6.8.7.5 *Grounding/earthing system*

In the marketplace, we will implement pipe earthing because it is economical and durable (Table 6.17).

6.8.7.6 *Cost of materials*

Table 6.17 Cost of materials for pump house

Particulars	*Quantity*	*Cost/Unit*	*Total cost (in ₹)*
Wiring and accessories			
PVC conduit (50 mm Dia.)	40 m	45/m	1,800
PVC conduit T and L bends	25 Nos.	5/pc	125
5 mm² PVC Cu wire	120 m	80/m	9,600
1 mm² PVC Cu wire	85 m	20/m	1,700
PVC switchboard	1 Nos	100/pc	100
2HP motor/water pump	5 Nos	15,000/unit	75,000
Motor mounting accessory	5 Sets	150/set	750
MCB	6 Nos	250/pc	1,500
6-Way DB	1 Nos	500/pc	500
Switch	2 Nos	55/pc	110
Ceiling roses	2 Nos	35/pc	70
Tube light	2 Nos	600/unit	1,200
GI earth pipe	1 Nos	1,000/unit	1,000
GI earthing strip	15 m	35/m	525

(Continued)

Table 6.17 (Continued) Cost of materials for pump house

Particulars	*Quantity*	*Cost/Unit*	*Total cost (in ₹)*
Earthing materials	—	1,500	1,500
Solar installation			
Solar panels	20 Nos.	15,500/unit	310,000
Solar inverter	1 Nos.	105,000/unit	105,000
Solar charge controller	1 Nos.	950/unit	950
Battery	6 Nos.	13,500/unit	81,000
Panel wires	25 m	15/m	375
Solar panel mounting	20 Nos.	2,500/unit	50,000
Total			642,805 ≈ 643,000

6.8.7.7 Total cost of installation

Table 6.18 Total cost of installation for pump house

Material cost	₹ 643,000
Labor charges (10% of material cost)	₹ 64,300
Super vision charges (2% of material cost)	₹ 12,860
Contingency charges (5% of material cost)	₹ 32,150
Grand total	₹ 752,310

Therefore, the total cost of electrical installation for the marketplace is around ₹4,85,000 (Table 6.18).

6.8.8 Estimation for auditorium

6.8.8.1 Installation plan diagram (Figure 6.16)

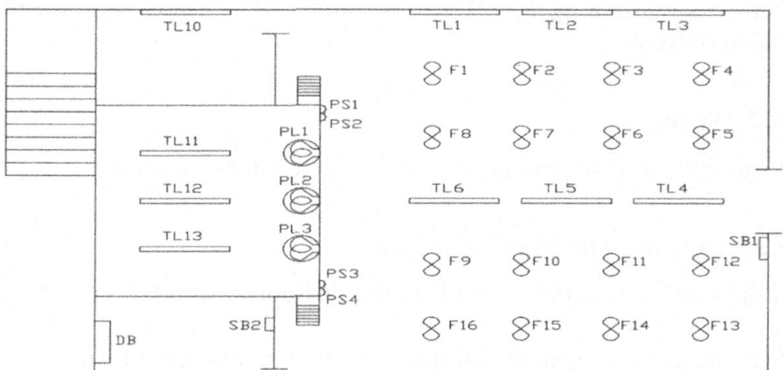

Figure 6.16 Auditorium installation plan.

Table 6.19 Load calculation for auditorium

Appliance	*Wattage (W)*	*Quantity*	*Total load (W)*
Tube lights	50	14	700
Power lights	200	3	600
Fans	100	16	1,600
Power sockets	1,000	4	4,000
Total lighting load			2,900
Total power load			4,000

6.8.8.2 Load calculation

Since, the total lighting load is approximately 2,900 W, we will be using four circuits. The first circuit will comprise all the tube lights, and the second will include all the power lights. The third and fourth circuits will be for fans (Table 6.19).

Additionally, the total power load is 4 kW. Therefore, we will be using two circuits. In total, we will need six circuits, and hence, a six-way distribution box will be used.

Circuit 1: 14 Tube Lights, Total Load = 700 W (Up to 800 W Permissible)
Circuit 2: 3 Power Lights, Total Load = 600 W (Up to 800 W Permissible)
Circuit 3: 8 Fans, Total Load = 800 W
Circuit 4: 8 Fans, Total Load = 800 W
Circuit 5: 2 Power Sockets, Total Load = 2 kW
Circuit 6: 2 Power Sockets, Total Load = 2 kW

Now, since the total number of lighting circuits is four and the permissible load in each circuit is 800 W, our total installed load capacity for lighting circuits is 3,200 W. However, the total connected load is only around 2,900 W. Thus, the remaining 300 W can be allocated as a provision for future extension of load.

6.8.8.3 Wiring

In the auditorium, we will use PVC concealed conduit wiring.

6.8.8.3.1 Calculation for length of conduit

The following assumption is made from the floor: ceiling = 5.5 m

$$\text{Stage height} = 1.5\,\text{m distribution box} = 2\,\text{m switch board} = 1.5\,\text{m}$$

Therefore, total length of conduit required is (Tables 6.20–6.22),

Table 6.20 Calculation for length of conduit for auditorium

Total horizontal run (in m)	*Total vertical run (in m)*	*Length of conduit required (in m)*
118	41	159
5% for wall crossing		7.9
10% for wall crossing		15.9
Total length of conduit required		182.8 ≈ 185

6.8.8.3.2 Calculation for size of conductor

Table 6.21 Calculation for size of conductor for auditorium

Circuit	*Load (W) (in watts)*	*Load current (W/230) (in Amps)*	*Size of conductor (Cu Wire) (in mm²)*
Circuit 1	800	3.5	2.5
Circuit 2	800	3.5	2.5
Circuit 3	800	3.5	2.5
Circuit 4	800	3.5	2.5
Circuit 5	2,000	8.6 ≈ 9	3.5
Circuit 6	2,000	8.6 ≈ 9	3.5

6.8.8.3.3 Calculation for length of conductor

Table 6.22 Calculation for length of conductor for auditorium

Location		*Distance (in m)*	*Size of conductor (in mm²)*	*Length of conductor (in m)*
From	*To*			
DB	SB1	31	2.5	31×2=62
	SB2	12		12×2=24
	PS1, PS2	17	3.5	17×3=51
	PS3, PS4	11		11×3=33
SB1	TL1, TL2, TL3	24	1	24×2=48
	F1, F2, F3, F4	24		24×2=48
	F5, F6, F7, F8	22		22×2=44
	TL4, TL5, TL6	12		12×2=24
	F9, F10, F11, F12	22		22×2=44
	F13, F14, F15, F16	24		24×2=48
	TL7, TL8, TL9	24		24×2=48
	TL10, TL14	46		46×2=92
	TL11, TL12, TL13	42		42×2=84

(Continued)

Table 6.22 (Continued) Calculation for length of conductor for auditorium

Location		*Distance (in m)*	*Size of conductor (in mm2)*	*Length of conductor (in m)*
From	*To*			
SB2	PL1	8	1	$8 \times 2 = 16$
	PL2	7		$7 \times 2 = 14$
	PL3	6		$6 \times 2 = 12$
Total length of 1 mm^2 conductor				522
Total length of 1 mm^2 conductor + 15% extra				$600.3 \approx 600$
Total length of 2.5 mm^2 conductor				86
Total length of 2.5 mm^2 conductor + 15% extra				$98.1 \approx 100$
Total length of 3.5 mm^2 conductor				84
Total length of 3.5 mm^2 conductor + 15% extra				$96.6 \approx 100$

6.8.9 Solar installation planning of the auditorium

In the auditorium, the total load is approximately 6,900 W. However, the installed load capacity is 7,200 W. Therefore, the design of the solar installation must accommodate the need for 7,200 W of energy. All the solar equipment must meet the criteria for serving the total load.

6.8.9.1 Solar panels

The solar panel we will be using is Luminous's mono-crystalline 380 W-24 V panel. To supply the load, we will need 19 such panels, which will provide a total load of 7,220 W.

6.8.9.2 Solar Inverter

Since our total load is about 7,200 W (7.2 kW), we will use Flin energy's hybrid inverter rated at 8 kW–24 V/240 V, keeping a provision for future extension of load.

6.8.9.3 Solar charge controller

The total load current of our house is (7,200/230=31.35 A), or approximately 35 A. Therefore, we will be using Luminous's MPPT-type solar charge controller with a 40 A rating, allowing for future load expansion.

6.8.9.4 Battery bank

Since the total load current is about 35 A and we are considering a battery run time of 24 hours at full load, we will need a battery bank with a capacity of 840 Ah.

To achieve this battery bank, we will use six Luminous lead acid batteries, each with a rating of 150 Ah-24 V, all connected in parallel.

6.8.9.5 Grounding/earthing system

In the auditorium, we will implement pipe earthing because of its cost-effectiveness and durability (Table 6.23).

6.8.9.6 Cost of materials

Table 6.23 Cost of materials for auditorium

Particulars	*Quantity*	*Cost/Unit*	*Total cost (in ₹)*
Wiring and accessories			
PVC conduit (50 mm Dia.)	185 m	45/m	8,325
PVC conduit T and L bends	70 Nos.	5/pc	350
3.5 mm² PVC Cu wire	100 m	56/m	5,600
2.5 mm² PVC Cu wire	100 m	45/m	4,500
1 mm² PVC Cu wire	600 m	20/m	12,000
PVC switchboard	2 Nos	100/pc	200
MCB	6 Nos	450/pc	2,700
Six-way DB	1 Nos	500/pc	500
Switch	33 Nos	55/pc	1,815
Ceiling roses	33 Nos	35/pc	1,155
Tube light	14 Nos	750/unit	10,500
Ceiling fan	16 Nos	1,500/set	24,000
Power socket with switch	7 sets	80/set	560
GI Earth pipe	1 Nos	1,000/unit	1,000
GI Earthing strip	10 m	35/m	350
Earthing materials	—	1,500	1,500
Solar installation			
Solar panels	19 Nos.	15,500/unit	294,500
Solar inverter	1 Nos.	105,000/unit	105,000
Solar charge controller	1 Nos.	950/unit	950
Battery	6 Nos.	13,500/unit	81,000
Panel wires	60 m	15/m	900
Solar panel mounting	19 Nos.	2,500/unit	47,500
Total			604,905 ≈ 605,000

6.8.9.7 Total cost of installation

The total cost of installation for the auditorium is ₹7,07,850, or approximately ₹7,10,000 (Table 6.24).

Table 6.24 Total cost of installation for auditorium

Material cost	*₹ 605,000*
Labor charges (10% of material cost)	₹ 60,500
Super vision charges (2% of material cost)	₹ 12,100
Contingency charges (5% of material cost)	₹ 30,250
Grand total	₹ 707,850

6.8.10 Estimation for solar street lighting system (SASSLS)

In this type of system, a single setup is provided for each solar street light, making it completely independent from other solar lights.

The system comprises a solar panel, a solar LED light, and a battery box. This system operates automatically through sensors, specifically photo-diodes. During the daytime, the solar panel charges the battery, and the light remains OFF. At night, when sunlight is not available, the light automatically turns ON. The light receives its power from the energy stored in the battery during the day. This is the most efficient solar street lighting system.

6.8.10.1 Various solar street lighting systems available

Table 6.25 Various solar street lighting systems available

Company	*Setup type*	*Specification*	*Price (₹)*
Havells	Without pole	50 W	9,700
Bajaj	Without pole	50 W	7,000
Luminous	With pole	70 W	17,500
Bajaj	Without pole	70 W	10,500
Luminous	Without pole	70 W	7,500

6.8.10.2 Installation planning

For the solar society, we will be using Luminous's With Pole, 70 W, stand-alone solar street lighting system (with reference to Table 6.25). The distance between two solar street lighting systems will be approximately 50 meters. Street intersections will be equipped with street lights. Based on our calculations, the total number of SASSLS will be 68. The placement of the SASSLS is shown in the 3D parametric view of the solar society (Tables 6.26–6.28).

6.8.10.3 Cost of materials

Table 6.26 Cost of materials for SASSLS

Particulars	*Quantity*	*Cost/unit*	*Total cost (in ₹)*
Solar street lighting setup	68 Nos.	17,500/setup	1,190,000
Pole mounting materials	68	2,000	136,000
GI Earth wire (8SWG)	(68×2) m	25/m	3,400
Total			1,329,400≈1,330,000

6.8.10.4 Total cost of installation

Table 6.27 Total cost of installation for SASSLS

Material cost	*₹ 1,330,000*
Labor charges (10% of material cost)	₹ 133,000
Super vision charges (2% of material cost)	₹ 26,600
Contingency charges (5% of material cost)	₹ 66,500
Grand total	₹ 1,556,100

The total cost of installation for the street lighting system will be ₹15,56,100, or approximately ₹15,60,000.

6.8.10.5 Total cost of installation for solar society

Table 6.28 Total cost of installation for solar society

Houses	*₹3,05,00,000*
Market place	₹4,85,000
Pump house	₹7,55,000
Auditorium	₹7,10,000
Street lighting system	₹15,60,000
Grand total	**₹3,40,10,000**
Grand total (in words)	**INR three crores, forty lakhs and ten thousand**

6.9 MAINTENANCE OF SOLAR SOCIETY

Maintenance can be referred to as the timely inspection and repair of equipment so that durability and sustainability can be obtained from equipment on a large scale. Everything needs maintenance, and so do electrical equipment. As the solar society is equipped with a large number of electrical devices, these also need maintenance to run properly. Proper inspection and maintenance of solar equipment should be done to obtain greater efficiency and long-term durability.

6.9.1 Solar panel maintenance

The solar panels are very durable devices. These panels need very little maintenance, and the cost is minimal. It includes inspection for dust and other material deposits on the panel, which may affect the efficiency of the panel. In case a layer of dust or other material is found to be deposited, it should be wiped off with water and a cloth.

6.9.2 Battery maintenance

As the batteries we are using here are lead acid type, one of the main components they contain is distilled water. Therefore, the water level should be checked once a month, and if the level is found to be low, it should be refilled soon. The cost of battery water is as low as ₹15 per liter, making it very inexpensive to maintain. Moreover, nowadays, lithium-ion batteries are being developed to resolve this problem. However, these batteries are currently very expensive, although the price is decreasing daily, and we can replace lead acid batteries with Li-ion batteries with future upgrades.

6.9.3 Solar inverter maintenance

The solar inverters are also static devices and require very little maintenance. An insulation resistance test can be performed periodically to check the insulation level and durability. The connections of the solar inverters should also be checked for any loose connections.

6.9.4 Wiring maintenance

Wiring maintenance is one of the most important factors. Such tests should be performed once a year. An insulation resistance test should be conducted. Boards should be inspected thoroughly for any loose connections, which may cause short circuits.

6.9.5 Grounding/earthing maintenance

The earthing resistance of earth pits should be checked yearly. The earthing resistance should be less than or equal to 1 Ω. If it is found to be higher, a mixture of salt and water should be added to the earthing pits.

6.10 FUTURE SCOPE OF WORK

As we have examined this work, it is undoubtedly clear that solar energy carries significant potential for meeting our energy demands. On the other hand, it is also observed that it has some drawbacks. The first major drawback is that solar energy is completely dependent on weather conditions. If the weather is not suitable for solar energy generation, the solar energy setup may not serve its purpose. The battery bank may provide backup to some extent. However, if unfavorable weather conditions persist for many days—which is highly uncertain—we may face a complete blackout.

Hence, to overcome this system, we may also use some other combination of non-conventional energy with the existing setup. In recent days, engineers have developed a hybrid system of non-conventional energy sources. In this type of system, two systems of non-conventional energy sources are merged together to serve a single load.

In this system, a power blackout is very rare, as it relies on two systems. This means that the failure of one system may be backed up by the other system as well. In our project, we may add a backup system to the existing solar setup. The most feasible setups are discussed further.

6.10.1 Wind with solar hybrid power setup

In this kind of system, wind turbines are used with the solar power system as a power backup. This setup is subject to geographical location and the availability of adequate wind speed in that area. The minimum wind speed required for such power plants to operate is around 25 m/s. These wind turbines are large and also require considerable space for installation. They are very costly and depend on the availability of wind. Like in the previous hydropower system, this system will also require a transmission line system for supply.

6.10.2 Biogas with solar hybrid power setup

Biogas power plants can be one of the most effective ways to provide a power backup to our society. This power plant mainly runs on organic and human waste. Biogas plants are not very expensive to set up and require less space compared to others. Although the efficiency of such plants is relatively low, they will definitely provide a sustainable power backup to our existing solar power setup.

6.11 CONCLUSION

From this chapter, we can conclude that the future is completely dependent on non-conventional sources. For that, solar energy is the most efficient solution at ground level. From small solar power setups to large solar harnessing gardens, solar power can be implemented on a very large scale. Although the installation cost is somewhat high, it has almost negligible maintenance costs. Additionally, the cost of solar power equipment is decreasing each day, and in the next few years, it will become highly affordable for all. Above all, the society we designed is a zero carbon emission society in terms of electrical power generation, thus protecting Mother Nature.

FURTHER READINGS

1. Boyer, C. B., & Merzbach, U. C. (2011). *A History of Mathematics*. John Wiley & Sons.
2. Gregory, A. (2015). *The Pythagoreans: Number and Numerology. Mathematicians and Their Gods: Interactions between Mathematics and Religious Beliefs*. Oxford University Press.
3. Osuji, U. (2021). A philosophical analysis of the role of numbers in ordering human activities and implications on epistemological discourses in Africa. *Journal of African Interdisciplinary Studies*, *5*(7), 148–162.
4. Visconti, S. (2020). *Numerology: Discover the Meaning Behind the Numbers in Your Life & Their Secrets to Success, Wealth, Relationships, Fortune Telling & Happiness*. Sofia Visconti.
5. Singh, P., Arora, K., Rathore, U. C., Yang, E., Joshi, G. P., & Son, K. C. Performance evaluation of grid-connected DFIG-based WECS with battery energy storage system under wind alterations using FOPID controller for RSC. *MDPI Journal of Mathematics*, 2023, 11(9), 2100. https://doi.org/10.3390/math11092100
6. Kovan, D. (2018). *Secrets of Numerology*. Ivy Press.
7. Ayonrinde, O. A., Stefatos, A., Miller, S., Richer, A., Nadkarni, P., She, J., ... & Mngoma, N. (2021). The salience and symbolism of numbers across cultural beliefs and practice. *International Review of Psychiatry*, *33*(1–2), 179–188.
8. Wilkie, J. E., & Bodenhausen, G. V. (2015). The numerology of gender: gendered perceptions of even and odd numbers. *Frontiers in Psychology*, *6*, 148470.
9. Lawrence, S. B. (2019). *The Big Book of Numerology: The Hidden Meaning of Numbers and Letters*. Weiser Books.
10. Bhatt, B,, Sharma, H., Arora, K., Joshi, G. P., & Shrestha, B, (2023). Levy flight-based improved grey wolf optimization: a solution for various engineering problems. *MDPI Journal of Mathematics*, 2023, 11(7), 1745. https://doi.org/10.3390/math11071745
11. Junior, L. A. (2019). *Numerology: Key to Being*. Editora Griot.
12. Singh, S. C. (2020). *Let the Numbers Guide You: The Spiritual Science of Numerology*. John Hunt Publishing.
13. Maria, F. (2021). *Empower Your Success with Numerology and Astrology*. BookRix.
14. Zhmud, L. (2019). From number symbolism to arithmology. In L. Schimmelpfennig & R. G. Kratz (Eds.), *Zahlen-und Buchstabensysteme im Dienste Religiöser Bildung* (pp. 25–45). Tübingen, Germany: Mohr Siebeck (Seraphim).

15. Connor, S. (2016). *Living by Numbers: In Defence of Quantity*. Reaktion Books.
16. Bawden, D., & Robinson, L. (2016). Information's magic numbers: the numerology of information science. In C. R. Sugimoto (Ed.), *Theories of Information and Scholarly Communication* (pp. 180–196). Berlin, Germany: De Gruyter Saur. https://doi.org/10.1515/9783110308464012.
17. Arora, K., Joshi, G. P., Ragab, M., Rawa, M., Milyani, A. H., Mansour, R. F., & Yang, E. (2022). Bilateral contract for load frequency and renewable energy sources using advanced controller. *Computers, Materials & Continua*, *73*(2), 3165–3180. https://doi.org/10.32604/cmc.2022.026966.
18. Scolnick, R. (2022). *The Witch's Book of Numbers: Enhance Your Magic with Numerology*. Red Wheel/Weiser.
19. Abduov, M. I., & Abduova, N. M. (2018). Function of holy numbers in lyrical and epic poetry. *Bulletin of the Karaganda university Philology Series*, *92*(4), 14–19.

Chapter 7

Recent innovations in energy storage systems

Denzel Christopher Makepa,
Anesu Nicholas Charamba, and
Hagreaves Kumba

7.1 INTRODUCTION

Across the globe, the energy landscape is undergoing a profound transformation, driven by the sustainable development goals to address climate change, reduce dependence on fossil fuels, and ensure sustainable energy access for all [1]. One approach to this transformation is the application of innovations in non-conventional energy sources, providing alternatives to traditional energy systems. Innovations in non-conventional energy sources are essential due to the depletion of conventional energy reserves like coal, environmental concerns, and the need for sustainable energy solutions [2].

This was highlighted in various research papers, which stated that the global energy economy heavily relies on fossil fuels and faces challenges such as environmental issues, depletion of natural reserves, and thermal pollution [2–4]. The role of energy storage systems in the energy transition for a sustainable energy future has been explored [5]. To address this gap, emerging technologies in unconventional energy sources are being investigated, such as battery technologies, thermal energy storage systems (TESS), mechanical energy storage systems, and other innovations, as shown in Table 7.1.

Table 7.1 highlights key innovations aimed at addressing the issues related to traditional energy sources and the need for sustainable energy solutions. In addition, these innovations are crucial in addressing the rise in global energy demand while minimizing environmental impacts and promoting a sustainable energy future.

The non-conventional energy sources cover a wide range of technologies, including wind, solar, geothermal, biomass, hydrogen, and ocean energy. These sources harness specific natural processes to generate power, each showcasing distinct advantages and challenges. Table 7.2 illustrates the innovations in non-conventional energy sources and their types.

From Table 7.2, the types of innovations across various non-conventional energy sources are highlighted, along with their role in meeting global energy needs while addressing environmental concerns and enhancing energy efficiency. As demonstrated by a few researchers, solar energy

 DOI: 10.1201/9781003542568-7

Table 7.1 Non-conventional sources of energy

Problem	*Innovation*	*Description*	*References*
Depletion of natural resources	Battery technologies, pumped hydro storage, flywheels	Application of energy storage systems designed to store energy from renewable sources efficiently.	[56]
Environmental concerns	Solar PV systems	Use of solar energy in houses, industries to reduce environmental risks and promote a sustainable environment.	[7–10]
Thermal pollution	Windmills, hybrid cooling systems, natural heat dissipation techniques	Technologies that capture excess heat from industrial processes, methods such as using natural bodies of water or geothermal cooling systems that leverage natural processes for heat management.	[11–13]
Need for sustainable energy solutions	Use of clean low-carbon energy sources	Shifting towards energy sources that produce minimal carbon emissions to combat climate change and ensure sustainability.	[4,14]

Table 7.2 Innovations in non-conventional energy sources

Type of source	*Description*	*Main examples*	*References*
Solar energy	Solar thermal systems, advances in photovoltaic cells, energy storage systems	Concentrated solar power Battery technologies Perovskite solar cells	[5,15]
Wind energy	Advances in wind turbine design with innovative blade designs, offshore wind technology	Floating wind farms, advanced horizontal-axis turbines, larger turbines	[16]
Geothermal energy	Improved geothermal systems, advanced exploration techniques like drilling	Enhanced geothermal systems, Geothermal energy storage systems, clime on heat power technology	[17,18]
Ocean energy	Advances in tidal and wave energy converters, research and development in materials and design	Wave energy converters, floating offshore wind turbines, tidal stream generators, salinity gradient energy converters	[19,20]

(Continued)

Table 7.2 (Continued) Innovations in non-conventional energy sources

Type of source	*Description*	*Main examples*	*References*
Hydrogen energy	Advances in the manufacturing, storage and the fuel cell technologies, enhanced hydrogen-based power generation	Solid state hydrogen storage, hydrogen powered cars, buses, trains	[21,22]
Waste to energy technologies	Research and development in waste conversion technologies	An aerobic digestion for municipal wastes, gasification, pyrolysis	[23,24]

harnesses the sun's abundant radiation through photovoltaic cells and concentrated solar power (CSP) systems, whereas wind energy captures the kinetic energy from the atmosphere using turbines [5,15]. Furthermore, geothermal energy utilizes the earth's internal heat, biomass energy converts organic materials into fuel energy, and ocean energy harnesses tidal and wave motions [2,17].

On the contrary, few studies have reported that recent innovations in these fields have substantially improved the efficiency, reliability, and scalability of non-conventional energy systems [14,25]. Research and development on advanced materials, improved storage solutions, and integrated smart grid technologies are reducing costs and boosting the feasibility of widespread adoption across the globe. An increasing body of literature shows that interdisciplinary research and collaborations are promoting novel approaches to overcome technical and economic barriers, supporting the integration of these renewable sources into existing energy infrastructures for a sustainable energy future [2,4].

This chapter delves into cutting-edge developments across various non-conventional energy sectors, highlighting key technological breakthroughs and their implications for sustainable development. The chapter will also explore the latest advancements in non-conventional energy technologies; for example, battery technologies and thermal energy storage (TES), examining their potential to reshape energy production, distribution, and consumption.

7.2 ADVANCEMENTS IN BATTERY TECHNOLOGIES

The advancements in battery technologies explore the latest developments in battery engineering and science. This highlights key developments in various types of batteries, including lithium-ion, flow batteries, and solid-state batteries. The evolution of battery manufacturing has seen a shift towards eco-friendliness and cost-efficiency, with a focus on dry electrode technologies to reduce energy consumption and environmental impacts. The emerging battery technologies offer a promising alternative to current lithium-ion batteries. These new battery technologies aim to address the limitations of lithium-ion batteries by improving performance, safety, cost, and durability.

7.2.1 Lithium-ion batteries

Research on lithium-ion batteries has primarily focused on enhancing energy density and safety. Recent advancements in lithium-ion battery technologies have centered on improving energy density, lifespan, and performance for electric vehicles (EVs) and portable electronics. Lithium-ion batteries have a higher energy density and are extensively used in EVs globally. However, lithium-ion batteries still have issues, as shown in Table 7.3, highlighting the need for further research and development.

A recent study by Wen et al. [26] highlighted methods to improve the energy storage and lifespan of lithium-ion batteries, also addressing challenges related to safety and battery management. The authors argued that battery management systems (BMSs) are crucial for enhancing battery safety, performance, and energy efficiency. Ongoing research indicates that progress in material and thermal characteristics is critical for battery efficiency and capacity. Key advancements and reviews of existing lithium-ion battery technologies, focusing on developments in cathode and anode materials, electrolyte formulations, and BMSs that have improved energy density, charging rates, and overall performance of EV batteries, have been explored [27]. Progress in material and thermal characteristics has been identified as vital for battery efficiency and capacity. Prior research suggests that lithium metal anodes are being researched and developed as a promising alternative to traditional lithium-ion batteries due to their potential for higher energy density [28]. Most early studies, as well as current work,

Table 7.3 Problems of lithium-ion batteries

Problem	*Key description*
Limited life-span	Capacity degradation due to repeated charging and discharging
Performance issues	Reduced efficiency in extreme weather conditions, hot or cold
Safety concerns	Risk of fires due to overheating
Energy density issues	Need for further research and improvement in energy density
Cost	Higher production costs
Environmental impacts	Challenges in recycling and disposal, leading to environmental pollution

focus on developing new materials for both cathode and anode designs to enhance the performance and lifespan of lithium-ion battery systems [29]. Additionally, advancements in nanotechnology, particularly in silicon and carbon nanotechnologies, have addressed challenges related to low power and energy density in lithium-ion batteries, showing promising solutions for enhancing battery performance [30]. Ongoing research highlights that advances in high-capacity electrode materials and integrated battery systems can improve the energy density and lifespan of lithium-ion batteries [31]. Research and development in lithium-ion batteries are crucial for modern energy storage. Some authors have driven the further development of lithium-ion batteries by looking at sustainability and recyclability, including material selection and green technology [27]. According to Lai et al. [32], the optimization of electrolytes and separators is essential for enhancing the performance and safety of lithium-ion batteries. Thus, the development of new liquid and solid electrolytes that can operate at higher voltages enhances battery performance. A recent study by Chen et al. [33] concluded that post-ion technologies, such as lithium-sulfur, offer higher energy density compared to current lithium-ion batteries. Advancements in lithium-ion battery technology are driven by the need for higher energy densities and longer lifespans. Innovations in electrode materials, electrolytes, and BMSs, along with the integration of AI, are key to achieving these goals. These collective efforts are paving the way for more efficient, sustainable, and long-lasting lithium-ion battery solutions.

7.2.2 Flow batteries

Research and development in flow batteries, particularly focusing on the development of vanadium and zinc–bromine flow batteries, has been a key area of innovation in energy storage technologies. Flow batteries, especially vanadium and zinc–bromine systems, are gaining significant attention for their potential in large-scale energy storage applications. These batteries are essential for integrating renewable energy sources (RES) into the grid, addressing issues of intermittency and stability for a sustainable energy future.

Several studies suggest that zinc–bromine flow batteries have emerged as a strong contender for applications in large-scale energy storage due to their low cost, high energy density, and environmentally sustainable profile [34,35]. Another study by Ulaganathan et al. [36] highlighted that zinc–bromine batteries exhibit voltaic efficiency, coulombic efficiency, cycle stability, and environmental sustainability. Research and development on these batteries demonstrate scalability and flexibility, making them suitable for both grid-scale and residential applications [34]. The development of new zinc–vanadium redox flow batteries represents a significant advancement in energy storage technology. Despite decades of research, recent studies have focused on improving cell structure and electrode design to enhance the performance of the batteries. Another review emphasized that flow batteries are crucial for renewable energy storage, with a need for improved power and energy densities, cost reductions, and structural design optimization for industrial applications [37]. More focused research on membrane costs, electrochemical kinetics, and electrode design is needed for the widespread commercialization and adoption of flow batteries [38]. For example, research has provided evidence that advancements in cell structure and electrodes significantly enhance the performance and practicality of zinc–bromine flow batteries for energy storage applications [35].

Key insights on the development of vanadium and zinc–bromine flow batteries:

- Scalability: Both technologies are scalable, but vanadium flow batteries are often preferred for very large-scale applications due to their longer cycle life [35].
- Cost: Zinc–bromine flow batteries generally have a lower initial cost, but the total cost of ownership may be higher due to maintenance requirements [39].
- Safety and Environmental Impact: Vanadium flow batteries are considered safer and have a lower environmental impact due to fewer concerns regarding toxic materials [40].

Vanadium and zinc–bromine flow batteries are promising technologies for large-scale energy storage, each with unique advantages and challenges. Vanadium flow batteries are noted for their efficiency and scalability, while zinc–bromine batteries offer high energy density and lower material costs. However, both systems face challenges related to cost, energy density, and environmental impact. The advancements in zinc–bromine flow batteries, such as improved cell structures and electrode materials, present potential benefits, such as enhanced performance and increased energy efficiency, but they also pose some challenges. The integration of new materials and designs can increase manufacturing costs and add complexity to the system, potentially leading to higher maintenance and reduced reliability. A number of questions regarding cost and safety remain to be addressed. In addition, the long-term stability of these improvements must be validated, and their environmental impacts should be carefully assessed. To fill this literature gap, these enhancements must outperform competing energy storage technologies to ensure a significant market advantage.

7.2.3 Emerging battery technologies

The increasing demand for efficient and cost-effective energy storage solutions has led to significant research into alternative battery technologies. Sodium-ion and lithium–sulfur batteries are two emerging technologies that offer potential advantages over traditional lithium-ion batteries.

These batteries provide superior energy density and low material costs, making them suitable for various applications across the globe.

Advanced research and development in lithium–sulfur batteries focus on enhancing performance. Recent studies have highlighted that incorporating hybrid carbon nanotubes and a nickel–zinc ferrite-coated separator can significantly improve high-rate capabilities and reduce polarization in these batteries [41]. Several authors have suggested research and advancement of all-solid-state lithium–sulfur batteries to address the challenges of liquid electrolyte-based batteries, emphasizing safety and energy density improvements [42]. To enhance battery performance, scientists have expanded their research to advancements in sulfur-based composite cathode design, separator modification, binder improvement, electrolyte optimization, and lithium metal protection [43,44].

A more comprehensive study by Ni [45] also highlights the importance of addressing challenges like the shuttle effect, reaction kinetics, and anodes in practical lithium–sulfur battery applications, proposing solutions such as advanced electrodes, electrolytes, and characterization strategies. Furthermore, another author reported that high-activity electrocatalysts have shown promising results in enhancing sulfur redox reaction kinetics and overall battery performance [46].

A recent study by Sun et al. [47] discussed the potential use of lithium–sulfur batteries as a superior alternative to lithium-ion batteries. The study indicated that these batteries offer a remarkably higher theoretical energy density due to the large capacity of sulfur. However, previous studies also noted some issues, such as the poor electronic conductivity of sulfur and problems with cycle stability [48]. The authors suggest that overcoming these technical barriers through material and engineering innovation could make lithium–sulfur batteries the top choice for future energy storage, outperforming current battery technologies such as lithium-ion. Therefore, with ongoing research and development efforts, lithium–sulfur batteries hold the promise of surpassing lithium-ion batteries and becoming the leading option for future energy storage applications. The application of these batteries offers increased potential energy density, reduced costs, and improved safety prospects, making them a frontrunner among next-generation batteries.

Sodium-ion and lithium–sulfur batteries represent promising advancements in energy storage technology, each with unique advantages and challenges. Sodium-ion batteries offer a cost-effective and abundant alternative to lithium-ion batteries, with significant progress in electrode materials and cell design. Room temperature sodium–sulfur batteries are particularly promising for large-scale applications, though they face challenges related to electrode stability and polysulfide management. Lithium–sulfur batteries, with their high energy density, are advancing through improvements in cathode design, electrolyte optimization, and anode protection.

Both technologies share common challenges, such as polysulfide dissolution and dendrite formation, which are being addressed through innovative strategies like functional separators and atomic layer deposition. Continued research and development in these areas are essential for realizing the full potential of these next-generation battery technologies.

7.3 INNOVATIONS IN SUPERCAPACITORS

7.3.1 Structure and function

Supercapacitors consist of two electrodes separated by an electrolyte. Unlike batteries, they store energy through the electrostatic separation of charge rather than through chemical reactions. This allows for rapid charging, with supercapacitors capable of being charged within seconds to minutes. Their high power density enables them to deliver large amounts of power quickly, and they have an extended cycle life, capable of millions of charge–discharge cycles without significant degradation [49,50].

7.3.2 Types of supercapacitors

Supercapacitors, a vital component in energy storage systems, primarily fall into two main categories: electric double-layer capacitors (EDLCs) and pseudocapacitors, each with distinct mechanisms for energy storage and unique applications.

EDLCs store energy through the formation of an electric double layer at the interface of the electrode and the electrolyte. When voltage is applied, ions from the electrolyte adhere to the surface of the electrode, creating a charged layer. The electrodes commonly utilize activated carbon due to its exceptional surface area, which can exceed 3,000 m^2/g [51]. This high surface area facilitates significant energy storage in a compact form, making EDLCs highly efficient and cost-effective. The choice of electrolyte, whether aqueous or organic, greatly influences performance characteristics such as voltage handling, conductivity, and temperature stability. Aqueous electrolytes, while limited to lower voltages, provide higher conductivity, whereas organic electrolytes enable higher energy densities and working voltages, making EDLCs suitable for a range of applications from consumer electronics to EVs [52].

In contrast, pseudocapacitors rely on rapid and reversible redox reactions occurring at the electrode surface to store energy. This process allows for a higher energy density than what is achievable with traditional EDLCs. Pseudocapacitors often incorporate materials such as conducting polymers, like polyaniline and polypyrrole, which enhance charge storage through electrochemical doping [53]. Additionally, metal oxides, including manganese dioxide (MnO_2) and ruthenium oxide (RuO_2), are commonly used due to their excellent charge storage capabilities. These materials enable pseudocapacitors to achieve both high capacitance and significant power density, making them ideal for applications requiring quick bursts of energy, such as regenerative braking systems in EVs or backup power for critical electronic devices [54].

7.3.3 Recent innovations in supercapacitor technology

Recent advancements in supercapacitor technology have led to groundbreaking innovations, notably in the area of graphene-based supercapacitors. Graphene, which consists of a single layer of carbon atoms arranged in a two-dimensional lattice, boasts remarkable properties, including exceptional electrical conductivity, high thermal conductivity, and an extensive surface area. These characteristics enable graphene supercapacitors to achieve energy densities increasingly comparable to those of conventional batteries [55]. The high energy density not only enhances performance but also contributes to lighter and more compact energy storage solutions. Moreover, the inherent flexibility of graphene allows for its integration into flexible electronics and wearable devices, paving the way for innovative applications such as smart textiles and portable gadgets that require reliable power sources.

Another significant innovation in supercapacitor technology is the emergence of hybrid supercapacitors, which effectively combine the attributes of traditional batteries and supercapacitors [56]. These hybrids utilize asymmetric electrodes, typically featuring a battery-type anode paired with a capacitor-type cathode, to capitalize on the strengths of both technologies. This configuration significantly enhances both energy density and power density, allowing these systems to operate over a broader voltage range. As a result, hybrid supercapacitors are particularly suitable for applications that demand higher voltage outputs, such as EVs, renewable energy systems, and advanced industrial machinery [57]. Hybrid designs provide a more versatile energy storage solution by overcoming some limitations of standard supercapacitors, addressing the growing demand for efficient and high-performing energy systems in various sectors.

7.4 THERMAL ENERGY STORAGE

7.4.1 Overview of TESS

The grid supply instability resulting from the rapid integration of variable RES creates a need for operational flexibility on various timescales. The surplus energy from RES during off-peak times can be converted into heat and stored in TESS for later use [58]. TES technologies store thermal energy in well-insulated materials by changing the materials' internal energies through heating, cooling, melting, or solidifying [59]. TESS plays an important role in demand-side management, particularly in buildings, by catering to the cooling, heating, and domestic hot water (DHW) needs through integration into ceilings, ventilation or PV systems, walls, and solar facades, etc. [60].

Integrating seasonal and short-term TESS can be used for simultaneous daily and seasonal peak shaving of heat demands. Seasonal TESS stores surplus heat energy and enables heat shifting when demand is at peak, thus eliminating energy imbalances across multiple time scales [58]. As buildings are the largest energy consumers in the built environment, TESS technologies are primarily adopted for DHW systems as well as heating, ventilation, and air conditioning systems [61]. Additionally, most of the TESS technologies provide heat energy to target systems rather than converting it into electricity due to the significant losses involved in the process [62]. Generally, the relationship between temperature and heat energy is given by Equation 7.1 [62].

$$T_{\text{energy}} = m * C * \Delta T \tag{7.1}$$

where C is specific heat capacity, T_{energy} is the amount of heat energy stored, m is the mass of storage body, and ΔT is change in temperature.

The application of TESS in new or retrofitted buildings enhances the performance and energy efficiency of current systems [60]. Overall, TESS technologies can lead to lower operational costs, increased efficiency and reliability, and reduced greenhouse gas emissions. For example, an estimated 400 million tons of CO2 emissions can be prevented in the buildings and industrial sectors by employing TESS [60]. Additionally, having large storage systems increases non-transitional hours with the grid, and the formation of an energy community results in capital savings and shorter payback periods compared to when each house has a separate storage system [63].

7.4.2 Characteristics of TES

These are indicators used to describe a storage system and are particularly important during the design and development stages. The costs, storage periods, and efficiency are critical in determining the profitability of the system [62].

- Storage Capacity: The amount of energy that can be stored in the system. This depends on the storage medium, the process being implemented, as well as charge and discharge times.
- Charge and Discharge Times: The amount of time required to charge or discharge the storage system.
- System Cost: Expressed in per kWh (capacity) or in per kW (power). The overall costs depend on factors such as operational and maintenance costs, required capital, and system lifespan.

- Efficiency: The ratio of supplied energy to the energy required for charging the system.
- Storage Period: Refers to how long the energy is stored or lasts. It can be expressed in hours, months, etc. [60].
- Maturity Level: The stage at which the technology is currently in terms of its development. These levels can be demonstration, developing, commercialized, or mature.

7.4.3 Types of TES technologies

Thermal storage systems can be classified as sensible heat storage (SHS) due to the temperature change that occurs within the material, thermo-chemical storage (TCES) when a chemical reaction occurs at the surface of the material, or latent heat storage (LHS) if there is a phase change without temperature change in a material during the charge and discharge processes [62]. SHS is a mature TES technology, whereas LHS and TCES are still under laboratory testing [58]. Classes of different TESS are depicted in Figure 7.1, and brief descriptions of some of them follow.

7.4.3.1 Sensible heat storage

Water, molten salt, sand, rocks, and molten aluminum are examples of storage mediums in SHS, with molten salt being the most commercially developed technology, such as in CSP plants. This is because salt is environmentally friendly, and the more it is heated, the higher the energy efficiency. The specific heat capacity of the medium is also important in designing a storage system, as low specific heat capacities require a larger storage unit [60].

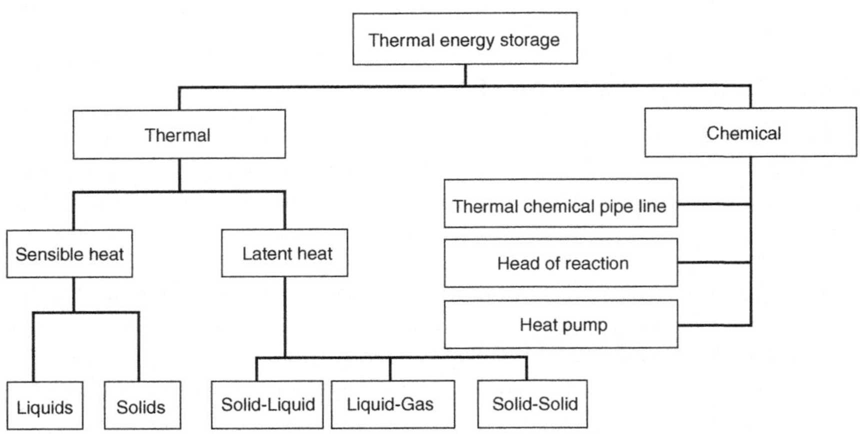

Figure 7.1 Classification of thermal energy storage systems [60].

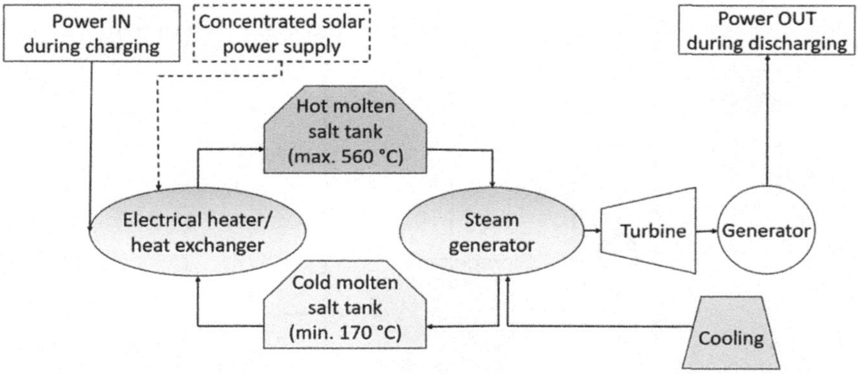

Figure 7.2 An illustration of a molten salt energy storage system [62].

7.4.3.1.1 Molten salt

During the charging process, electrical heaters heat the salt until it becomes molten, typically around 600°C. Molten salt is stored in insulated hot storage tanks, with temperature losses around 1°C, for about 8–16 hours. During discharge, the hot molten salt heats the water to produce steam which is used to turn generator turbines. Used molten salt is stored in a cold storage tank, waiting to be reheated during recharge [62]. Due to higher energy losses, thermal PV cells are set to substitute turbines in these systems. Figure 7.2 shows a typical molten salt energy storage system.

7.4.3.1.2 Hot water storage tank

On the other hand, water is the cheapest medium, has a high specific heat, and is commonly used in residential and industrial applications, but has a relatively lower storage capacity since it cannot be heated to as high temperatures [60]. A hot water storage tank is cost-effective, and its efficiency can be improved by enhanced thermal insulation, e.g., evacuated super-insulation. For an unstratified tank, the energy balance is given by [60]:

$$mC\frac{dt_s}{d\tau} = Q_U - Q_L - U_s A_s\left(t_f - t_a\right) \tag{7.2}$$

where m is the mass of the storage body, $\frac{dt_s}{d\tau}$ is the temperature range with time, Q_U and Q_L are the rates at which energy is added or removed, U_s is the storage tank's heat loss coefficient, A_s is its surface area and t_f, and t_a are final and ambient temperatures of the tank.

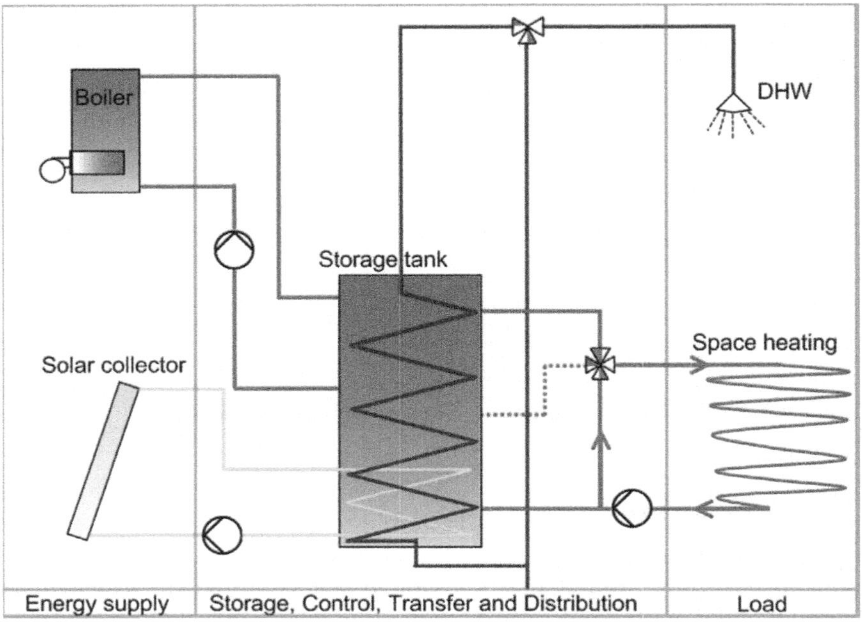

Figure 7.3 Example of a combi-system [60].

To predict the performance of the system over time, Equation 7.3 is used to approximate the temperature of the water storage as a function of time.

$$t_s = t_i + \frac{\Delta\tau}{mC}\left[Q_U - Q_L - U_s A_s\left(t_f - t_a\right)\right] \tag{7.3}$$

Hot water storage can be used in DHW supplies, but when DHW supplies are integrated with building heating systems, they form combi-systems. An illustration of a combi-system comprising two heat sources, a solar collector and a boiler, and two heat sinks, DHW and space heating, is shown in Figure 7.3.

7.4.3.1.3 Underground TES

Underground TES systems are suitable for both heat and cold storage. A heat transfer fluid is pumped through underground pipes to add or extract heat energy from the storage system. The rate of heat transfer and the area of the pipes determine the charge and discharge rates. Three common configurations used in this storage technology are boreholes, aquifers, and pit storages. Boreholes are used in conjunction with heat pumps, whereas in aquifers, heat is exchanged through pumping water into or out of the ground. Aquifers are commonly applied for cold storage. However, since they cannot be insulated, heated water can chemically react with ground materials, leading to reduced efficiency. Pit storages, on the other hand, have relatively high investment costs, which limits their adoption.

7.4.3.2 LHS

LHS uses phase change materials (PCMs) to provide better storage capacity [59]. These materials can absorb or release heat energy during a change in their physical state at an almost constant temperature [60]. The phase changes can be solid to solid, solid to gas, solid to liquid, or liquid to gas. An LHS system comprises a PCM, a PCM container, and a heat exchange surface. Differential thermal analysis and differential scanning calorimetry are used to measure the latent heat of fusion and the melting temperatures of PCMs, in which samples and reference materials are heated, and their temperature differences, proportional to their heat flow differences, are used to plot a differential scanning calorimetry curve [60]. The heat energy storage capacity of a PCM-based LHS system is determined by Equations 7.4 and 7.5.

$$Q = \int_{t_i}^{t_m} mCdt + mf\Delta q + \int_{t_m}^{t_f} mCdt \tag{7.4}$$

$$Q = m\left[C_s\left(t_m - t_i\right) + f\Delta q + C_l\left(t_f - t_m\right)\right] \tag{7.5}$$

where m is the mass of the PCM material, f is the melt fraction, Δq is the latent heat of fusion, C_s and C_l are specific heat capacities of the solid and liquid phases, respectively.

Considering negligible axial conduction and bed heat losses, as well as a low Biot number, the energy balance in the material is given by Equation 7.6, and the energy balance on the fluid is given by Equation 7.7 [64].

$$\frac{\partial u}{\partial t} = \frac{\lambda_s}{\rho_s}\frac{\partial^2 t_s}{\partial x^2} + \frac{UP}{\rho_s A}\left(t_f - t_s\right) \tag{7.6}$$

$$\frac{\partial t_f}{\partial t} + \frac{m}{\rho_f A_f}\frac{\partial t_f}{\partial x} = \frac{UP}{\rho_f A_f C_f}\left(t_s - t_f\right) \tag{7.7}$$

where U is the heat transfer coefficient between the PCM and fluid, λ_s is the thermal conductivity, t_f is the fluid temperature, u is the specific internal temperature, ρ is the density, and A_f is the flow area.

7.4.3.2.1 Liquid air storages

In liquid air storage systems, air is cooled to about –196°C to liquefy it and store liquid air in unpressurized insulated vessels. In the charging process, the air is cleaned, compressed, cooled to liquefaction, and stored in an insulated vessel. Discharging is accomplished by pumping the liquid air, reheating it, and allowing the gas produced to expand, thus doing work on the turbines to produce electricity [62]. Since around 20% of the total energy is lost during the charging process, it is important to develop efficient ways of utilizing waste heat and waste cold. For example, 700 L of ambient air can be liquefied to 1 L, and when reheated during discharge, this liquid air converts into gas at a volume 700 times greater to drive a turbine or piston [62].

7.4.3.2.2 PCMs in buildings' active systems

Similar to integrating solar air collectors into building walls, PCMs can be integrated into ventilated facades. Nano-enhanced PCMs can also be included in solar buildings to improve the total thermal mass, thereby reducing the difference between energy supply and demand. An illustration of the working of PCM in a ventilated facade is shown in Figure 7.4 [60].

7.4.3.3 TCES

This TES technology involves the conversion of thermal energy into chemical energy and has the highest energy density compared to LHS and SHS. To store energy, a compound is heated resulting in two products: a solid material and a gaseous material. These products can be stored separately,

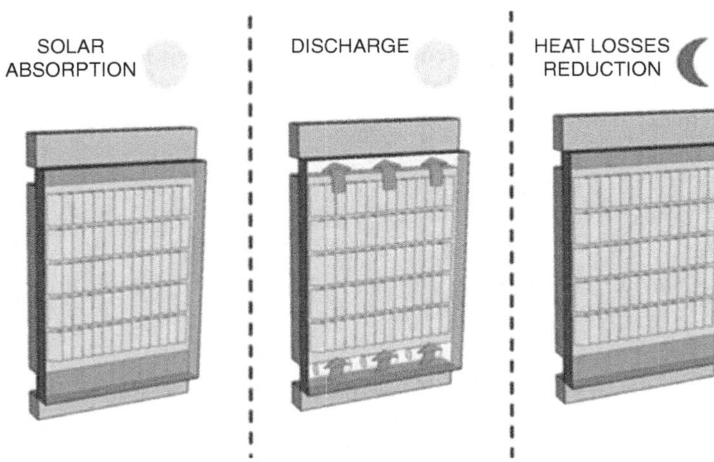

Figure 7.4 Working of a ventilated façade with PCM [60].

and during the discharge process, they undergo a chemical reaction under suitable operating temperature and pressure to produce energy [60]. Equation 7.8 illustrates this process [65].

$$n_{AB} \cdot AB_{(s)} + \Delta H \Leftrightarrow n_A \cdot A_{(s)} + n\ _B \cdot B_{(g)} \tag{7.8}$$

where $AB_{(s)}$ is a solid compound, ΔH is heat absorbed, $A_{(s)}$ is solid material, and $B_{(g)}$ is gaseous material.

TCES can be through chemical reactions or adsorption, with sorption processes enabling thermal energy transportation. On the other hand, chemical reactions can operate above 350 °C unlike sorption processes [60]. Examples of TCES systems include metal oxides, metallic carbonates, metal hydroxides, and metal hydrides. To store thermal energy, metal hydrides decompose and absorb heat, forming gaseous hydrogen and solid metals. Magnesium (Mg) hydrides are commonly used; however, their large-scale application is hindered by high operating temperatures and pressures [65]. Decomposition temperatures of metal hydroxides are in the range 300°C–700°C. Among alkaline earth metal hydroxides, Ca (OH_2) and Mg (OH_2) are common due to their non-toxicity and being relatively cheaper. However, their applications are limited by low thermal conductivities. On the other hand, metallic carbonates are suitable for applications in CSP plants as they can be operated at high temperatures [65].

7.4.3.4 Other emerging technologies

7.4.3.4.1 Sand batteries

Sand batteries are a new technology still in the developing stage. They use low-grade sand, which can be charged by RES such as solar or wind. The sand is heated up to 500°C by resistive heating. A heat exchanger circulates hot air through the sand, and the sand can be maintained at the same temperature for months [66]. During discharge, the heat energy is released and used to heat water that is supplied to heating systems. The major drawback is the drastic decrease in efficiency when converting the stored energy into electricity [65].

7.5 MECHANICAL ENERGY STORAGE SYSTEMS

Mechanical energy storage systems utilize physical processes to store energy, primarily through kinetic or potential energy. They are one of the oldest forms of energy storage and have been significantly enhanced by recent technological innovations.

7.5.1 Pumped hydro storage

Pumped hydro storage (PHS) is the most established and widely used form of energy storage globally. It operates by moving water between two reservoirs situated at different elevations. During periods of low energy demand, excess electricity is used to pump water from the lower reservoir to the upper one. When energy demand increases, the stored water is released back to the lower reservoir, passing through turbines to generate electricity [67].

Recent advances in turbine technology and hydraulic systems have enhanced the round-trip efficiency of PHS. Some modern facilities achieve efficiencies of over 80% due to improvements in materials and design [68]. Additionally, traditional PHS systems require significant geographical features, which limits their installation. Recent innovations have led to the development of smaller-scale pumped hydro systems that can be deployed in a wider variety of locations, including urban environments [69]. Furthermore, integrating pumped hydro with other renewable technologies, such as wind and solar, allows for better management of intermittent energy sources. Hybrid PHS systems can store energy from renewables while also providing grid stability.

7.5.2 Flywheel energy storage

Flywheel energy storage systems store energy in the form of kinetic energy through a rotating mass (flywheel). When energy is supplied, the flywheel accelerates, storing energy as rotational kinetic energy. Conversely, when energy is needed, the flywheel's kinetic energy is converted back into electricity [70].

Recent innovations in flywheel technology include the development of magnetic bearing systems, advanced materials, and integrated control systems. Innovations in magnetic bearing technology reduce friction, allowing flywheels to spin at higher speeds with minimal wear. This results in longer lifecycles and increased efficiency since friction losses are significantly minimized [71]. The development of carbon fiber and composite materials has led to lighter, stronger flywheels capable of operating at extremely high speeds. These materials improve energy density and overall system performance [71]. Modern flywheel systems feature advanced control algorithms that optimize charge and discharge cycles, allowing for rapid response times and more effective integration with grid management systems [72].

7.6 HYDROGEN STORAGE AND FUEL CELLS

Hydrogen is gaining recognition as a versatile energy carrier with significant potential for facilitating the transition to a sustainable energy system. It can be produced from various sources, including renewable energy through electrolysis, and used in diverse applications ranging from transportation to grid energy management. A critical aspect of utilizing hydrogen efficiently lies in effective storage methods and the development of robust fuel cell technologies [73].

7.6.1 Hydrogen storage technologies

Hydrogen storage can be categorized into three primary methods: compressed gas, liquid hydrogen, and solid-state storage. Each method presents distinct advantages and challenges regarding energy density, safety, and cost. Compressed Gas Storage involves storing hydrogen as a high-pressure gas in specially designed cylinders or tanks. This method is widely used due to its relatively simple infrastructure and ability to handle large quantities of hydrogen [74]. Recent innovations in composite materials have produced lighter and stronger tanks that can withstand higher pressures, thereby increasing storage efficiency [75].

Liquid hydrogen storage requires cooling hydrogen to extremely low temperatures, approximately –253 °C. While liquid hydrogen has a higher energy density compared to gaseous hydrogen, the energy required for liquefaction and the insulation needed to maintain low temperatures present significant challenges [76]. Innovations in cryogenic technologies and materials are being developed to enhance the efficiency of liquid hydrogen storage systems [77].

Solid-state storage offers promising alternatives through the use of materials that can absorb and release hydrogen. Metal hydrides are the most commonly explored materials in this category. They can store hydrogen at lower pressures and ambient temperatures compared to gaseous forms, improving safety and logistical convenience [78]. Ongoing research is focused on discovering new materials with higher hydrogen capacity and faster kinetics for absorption and release processes [79].

7.6.2 Fuel cells

Fuel cells are devices that convert the chemical energy of hydrogen directly into electricity through electrochemical reactions, without combustion, making them an attractive option for clean energy generation. The most common type is the proton exchange membrane fuel cell, which operates at relatively low temperatures and is suitable for applications such as vehicles and portable power systems [80].

Recent advances in fuel cell technology focus on improving efficiency, reducing costs, and enhancing durability. Innovations in catalyst materials, particularly the development of non-platinum catalysts, aim to lower dependence on precious metals, thus making fuel cells more economically viable for widespread adoption [81]. Furthermore, integrating fuel cells with hydrogen storage systems can create a seamless energy solution, particularly in transportation. Hydrogen fuel cell vehicles represent a growing segment of the automotive market, offering longer ranges and rapid refueling times compared to battery EVs. Automakers are increasingly investing in hydrogen technologies, supported by government initiatives promoting clean transportation.

7.7 INTEGRATION AND SMART GRID TECHNOLOGIES

7.7.1 Grid integration and smart grid technologies

With primary energy consumption projected to increase by 48% by 2040 and considering the intermittent nature of RES, the urgent development and integration of sustainable energy storage systems (ESS) has been recognized in recent years [60]. The transition of conventional power grids into green and sustainable smart grids results in unstable and unpredictable power systems due to the high penetration of variable renewables, hence the need to store surplus energy to provide ancillary services [82]. Smart grids incorporate distributed generation and demand response, while power scheduling and energy optimization are key features [83]. Providing system inertia, black start capabilities, and duck curve mitigation enables the wide deployment of ESS in such grids [84].

ESS technologies are being adopted to ensure the resilience and reliability of renewable energy systems during adverse weather conditions. Pumped-hydro energy storage occupies 96% of the total ESS installed globally, and lithium-ion battery energy storage systems constitute two-thirds of the global battery market, with the stationary BESS market predicted to grow to 1,095 GW/2,850 GWh by 2040 due to growing demand in EVs and stationary applications [84].

ESSs can be grid-tied for various purposes. Some applications include community and/or distributed ESSs, behind-the-meter energy, ancillary services, and energy arbitrage [85]. In smart grids, an ESS can be employed to provide black start services such as load balancing and voltage and frequency control. Black start is a process to restore an entire part of a power system after it has been shut down or destabilized by an unexpected event [86]. Energy storage enables time shifting by storing energy during off-peak times and using it when demand is high. Energy arbitrage is achieved in the same way by selling stored energy when electricity prices are high during peak demand [86]. In demand response programs, consumers update load patterns based on financial incentives offered by grid operators [83]. ESSs are also employed as standby reserves to provide uninterruptible power supplies in case of emergencies or, for example, in data centers. ESSs generate revenue through price arbitrage and grid services [86]. For inclusion in electricity markets, BESSs can be used for primary frequency regulation to respond to supply–demand imbalances, in most cases, using droop control (adjustment of generation active and reactive powers relative to frequency and voltage variations). Such systems are controlled by Equation 7.9.

$$\frac{df}{dt} = \frac{f_{\text{ref}}}{H_{\text{BESS}} + 2\sum H_n} \overbrace{(P_G - P_L)}^{\Delta P_d} \tag{7.9}$$

where H_n is system inertia from all rotating machines, ΔP_d is the change in power demand, P_G is generated power, P_L is load power demand, H_{BESS} is an inertia constant for BESS, and f_{ref} is nominal frequency [84].

Integration of solar PV and ESS in buildings has been increasing in recent years, hence the need for optimal control of these systems together with existing building loads to enhance energy efficiency and load flexibility [85]. An illustration of the integration of an ESS and solar PV into the grid for a single building is shown in Figure 7.5. Depending on the type, ESSs can be integrated into small-scale plants, individual households or energy communities (a group of households), and large-scale plants. ESSs are adopted in energy communities to enhance their self-consumption rates. Additionally, households can either feed into or consume from the ESS in the LV grid but only up to certain technically specified limits [82].

With the increased penetration of RES into the grid, the operational modes of ESSs are becoming more complex; hence, the relationship between these modes and system disturbances such as charge and/or discharge power must be optimized for the practically feasible real-world application of ESSs [85]. Demand side management is usually applied to regulate consumption based on instantaneous energy generation. In demand side management for energy communities, when there is surplus energy in the LV grid, the

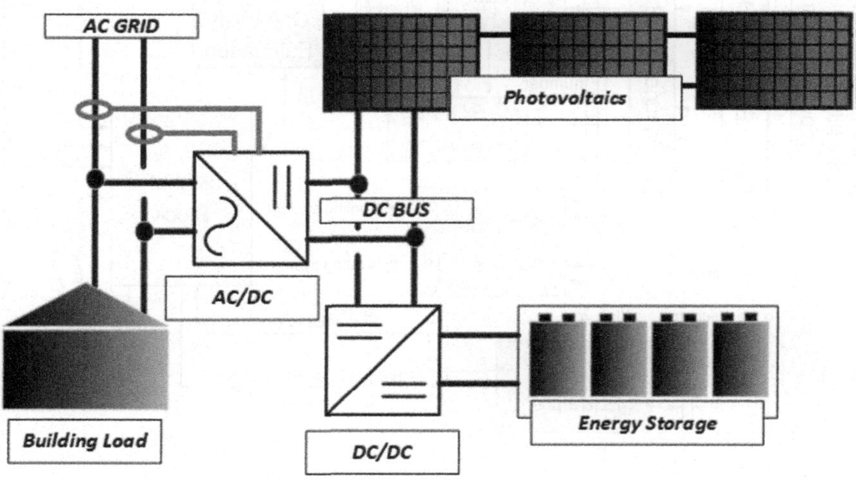

Figure 7.5 An example of integrating an ESS into the grid [85].

distribution system operator can request additional consumption. The load is shifted, and depending on its amount, financial incentives are then provided to the households by the distribution system operator [82].

The generation of excess PV energy in LV distribution systems can result in reverse power flows, leading to voltage rise challenges [87]. For optimal operation of integrated ESSs, optimization methods must be enhanced and employed to realize characteristics such as fast charge/discharge capability, reduced system interruptions, EV charging support, reductions in system operation and maintenance costs, a reduction in carbon footprint, and maximum energy production [85]. Methods that can be used include dynamic programming, linear programming, stochastic optimization, or machine learning algorithms such as reinforcement learning and long short-term memory, etc. [85]. To effectively control and prolong the life of grid-tied BESS, a BMS is required to safely operate the battery pack through cell monitoring, thermal management, and state of health/state of charge estimation [84]. A block diagram of a typical BMS is illustrated in Figure 7.6.

In grid-connected wind energy conversion systems, grid-side faults or variations in wind speeds can result in DC-link and grid voltage variations, leading to wind turbine trip-offs. This issue can be mitigated by effective control of active and reactive powers [88]. Active power storage systems such as BESS, superconducting magnetic storage, and supercapacitor energy storage can be implemented, as they can regulate active and reactive powers under many operating conditions. Supercapacitor energy storage provides

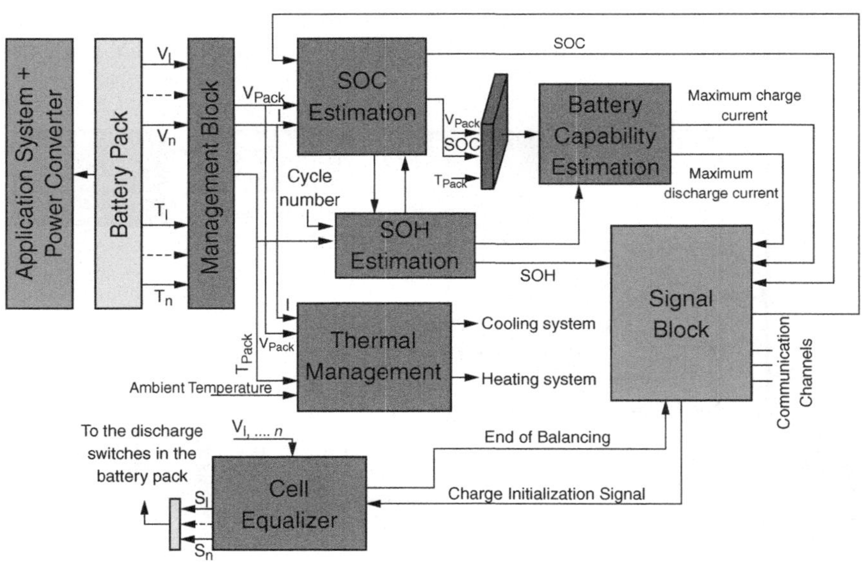

Figure 7.6 A block diagram of a typical BMS [84].

energy instantaneously and absorbs low to medium frequency variations; however, it lacks fault ride-through capability [88]. A hybrid ESS is advantageous because a high energy density ESS can store surplus energy to prevent reverse power flows, while a complementary ESS with high cycling capabilities can compensate for active power changes, thereby eliminating voltage fluctuations [87].

To size an ESS, the desired time frame for the power profile and data resolution are considered first; however, it is recommended to use high-resolution data and the entire frequency spectrum of the power profile [87]. In BESS, it is recommended to consider battery characteristics and load uncertainty in energy distribution and charge/discharge scheduling [85].

By using the most recurring pattern to represent the desired time series, an ESS can provide the necessary grid services. A pattern recognition algorithm, such as motif discovery, can be well-suited for this application among other algorithms [87]. Large datasets increase the computation time for the sizing process, and various methods are being developed to address this, such as converting high-resolution data into lower resolution or using one random power profile to represent certain time series, along with applying clustering algorithms like fuzzy C-means and K-means [87].

To accelerate the deployment of grid-tied ESSs, it is essential to develop reliability models, energy management, and analysis tools for long-term planning and real-time operation of such systems. In the case of BESSs, machine learning algorithms must be adopted to enhance state of charge/state of health estimation methods, as well as fire diagnostic and suppression technologies embedded in BMSs [84]. Advanced sensors can also be integrated into these BMSs to monitor current flow, thermal conditions, and other factors in real-time, thus improving reliability. Additionally, reconfigurable BESS can be implemented to avoid system failures when a failure occurs in one battery due to the fixed interconnectedness of most current BESSs [84].

7.8 CONCLUSIONS

Recent innovations in energy storage systems are pivotal to the advancement of renewable energy technologies and the broader transition towards sustainable energy solutions. This chapter has explored a diverse range of storage mechanisms, from batteries and supercapacitors to thermal and mechanical systems, each with unique advantages that cater to different energy demands and applications. As we face the challenges posed by intermittent energy sources like solar and wind, these innovations provide the necessary resilience and flexibility. Technologies such as lithium-ion batteries and hydrogen fuel cells are being refined to enhance efficiency, reduce costs, and improve sustainability. Moreover, the integration of energy storage systems with smart grid technologies presents a promising avenue for optimizing energy management, enabling real-time adjustments that enhance grid stability and responsiveness. Looking ahead, continued investment in research and development is essential to overcome existing barriers, such as high costs and the need for extensive infrastructure. As the global energy landscape evolves, energy storage systems will play an indispensable role in realizing a cleaner, more efficient, and resilient energy future, ensuring that we can effectively harness and utilize renewable resources to meet the demands of a sustainable world.

REFERENCES

1. B. Obuobi, F. Awuah, E. Nketiah, G. Adu-Gyamfi, V. Shi, and G. Hu, 'The dynamics of green innovation, environmental policy and energy structure for environmental sustainability; Evidence from AfCFTA countries', *Renew. Sustain. Energy Rev.*, vol. 197, p. 114409, Jun. 2024. https://doi.org/10.1016/j.rser.2024.114409
2. K. A. Khan, S. M. M. Manir, S. Islam, S. Jahan, L. Hassan, and M. H. Ali, 'Studies on nonconventional energy sources for electricity generation', *Int. J, Adv, Res, Innovative Ideas Educ.*, vol. 4, no. 4, pp. 229–244, 2018.
3. O. Q. F. Araújo, and J. L. De Medeiros, 'How is the transition away from fossil fuels doing, and how will the low-carbon future unfold?', *Clean Technol. Environ. Policy*, vol. 23, no. 5, pp. 1385–1388, Jul. 2021. https://doi.org/10.1007/s10098-021-02123-x
4. R. York, and S. E. Bell, 'Energy transitions or additions?' *Energy Res. Soc. Sci.*, vol. 51, pp. 40–43, May 2019. https://doi.org/10.1016/j.erss.2019.01.008
5. A. Kalair, N. Abas, M. S. Saleem, A. R. Kalair, and N. Khan, 'Role of energy storage systems in energy transition from fossil fuels to renewables', *Energy Storage*, vol. 3, no. 1, p. e135, Feb. 2021. https://doi.org/10.1002/est2.135
6. A. Blakers, M. Stocks, B. Lu, and C. Cheng, 'A review of pumped hydro energy storage', *Prog. Energy*, vol. 3, no. 2, p. 022003, Apr. 2021. https://doi.org/10.1088/2516-1083/abeb5b
7. H. Kumba, O. A. Olanrewaju, and R. Pasipamire, 'Integration of renewable energy technologies for sustainable development in South Africa: a focus on grid-connected PV systems', *Energies*, vol. 17, no. 12, p. 2823, Jan. 2024. https://doi.org/10.3390/en17122823
8. C. Breyer, D. Bogdanov, A. Gulagi, A. Aghahosseini, L. S. Barbosa, O. Koskinen, M. Barasa, U. Caldera, S. Afanasyeva, M. Child and J. Farfan, 'On the role of solar photovoltaics in global energy transition scenarios', *Prog. Photovolt. Res. Appl.*, vol. 25, no. 8, pp. 727–745, Aug. 2017. https://doi.org/10.1002/pip.2885
9. F. J. M. M. Nijsse, J. F. Mercure, N. Ameli, F. Larosa, S. Kothari, J. Rickman, P. Vercoulen and H. Pollitt, 'The momentum of the solar energy transition', *Nat. Commun.*, vol. 14, no. 1, p. 6542, Oct. 2023. https://doi.org/10.1038/s41467-023-41971-7
10. W. S. Ebhota, and T.-C. Jen, 'Fossil fuels environmental challenges and the role of solar photovoltaic technology advances in fast tracking hybrid renewable energy system', *Int. J. Precis. Eng. Manuf. Green Technol.*, vol. 7, no. 1, pp. 97–117, Jan. 2020. https://doi.org/10.1007/s40684-019-00101-9

11. T. Galimova, M. Ram, and C. Breyer, 'Mitigation of air pollution and corresponding impacts during a global energy transition towards 100% renewable energy system by 2050', *Energy Rep.*, vol. 8, pp. 14124–14143, Nov. 2022. https://doi.org/10.1016/j.egyr.2022.10.343
12. M. Child, O. Koskinen, L. Linnanen, and C. Breyer, 'Sustainability guardrails for energy scenarios of the global energy transition', *Renew. Sustain. Energy Rev.*, vol. 91, pp. 321–334, Aug. 2018. https://doi.org/10.1016/j.rser.2018.03.079
13. H. Mikulčić, J. Baleta, J. J. Klemeš, and X. Wang, 'Energy transition and the role of system integration of the energy, water and environmental systems', *J. Clean. Prod.*, vol. 292, p. 126027, Apr. 2021. https://doi.org/10.1016/j.jclepro.2021.126027
14. A. Ştefania Chenic, A. I. Cretu, A. Burlacu, N. Moroianu, D. Vîrjan, and D. Huru, 'Logical analysis on the strategy for a sustainable transition of the world to green energy—2050. Smart cities and villages coupled to renewable energy sources with low carbon footprint', *Sustainability*, vol. 14, no. 14, p. 8622, Jul. 2022. https://doi.org/10.3390/su14148622
15. N. Suresh Kumar and K. Chandra Babu Naidu, 'A review on perovskite solar cells (PSCs), materials and applications', *J. Materiomics*, vol. 7, no. 5, pp. 940–956, Sep. 2021. https://doi.org/10.1016/j.jmat.2021.04.002
16. J. Li, G. Wang, Z. Li, S. Yang, W. T. Chong, and X. Xiang, 'A review on development of offshore wind energy conversion system', *Int. J. Energy Res.*, vol. 44, no. 12, pp. 9283–9297, Oct. 2020. https://doi.org/10.1002/er.5751
17. N. Bartels, G. Bussmann, and R. Ignacy, 'Geothermal energy in the context of the energy transition process', in *2018 International IEEE Conference and Workshop in Óbuda on Electrical and Power Engineering (CANDO-EPE),* Nov. 2018, pp. 000103–000108. https://doi.org/10.1109/CANDO-EPE.2018.8601144
18. R. Moeck, R. Bracke, and J. Weber, 'The energy transition from fossil fuels to geothermal energy – a German case study', In *Proceedings World Geothermal Congress*, Beppu–Morioka, Japan, 2000, p. 1.
19. P. Chen and D. Wu, 'A review of hybrid wave-tidal energy conversion technology', *Ocean Eng.*, vol. 303, p. 117684, Jul. 2024. https://doi.org/10.1016/j.oceaneng.2024.117684
20. M. Z. A. Khan, H. A. Khan, and M. Aziz, 'Harvesting energy from ocean: technologies and perspectives', *Energies*, vol. 15, no. 9, p. 3456, May 2022. https://doi.org/10.3390/en15093456
21. A. Kovač, M. Paranos, and D. Marciuš, 'Hydrogen in energy transition: a review', *Int. J. Hydrog. Energy*, vol. 46, no. 16, pp. 10016–10035, Mar. 2021. https://doi.org/10.1016/j.ijhydene.2020.11.256
22. M. Yue, H. Lambert, E. Pahon, R. Roche, S. Jemei, and D. Hissel, 'Hydrogen energy systems: a critical review of technologies, applications, trends and challenges', *Renew. Sustain. Energy Rev.*, vol. 146, p. 111180, Aug. 2021. https://doi.org/10.1016/j.rser.2021.111180
23. A. R. Kalair, M. Seyedmahmoudian, A. Stojcevski, N. Abas, and N. Khan, 'Waste to energy conversion for a sustainable future', *Heliyon*, vol. 7, no. 10, p. e08155, Oct. 2021. https://doi.org/10.1016/j.heliyon.2021.e08155
24. W. Foster, U. Azimov, P. Gauthier-Maradei, L. C. Molano, M. Combrinck, J. Munoz, J. J. Esteves, and L. Patino, 'Waste-to-energy conversion technologies in the UK: processes and barriers – a review', *Renew. Sustain. Energy Rev.*, vol. 135, p. 110226, Jan. 2021. https://doi.org/10.1016/j.rser.2020.110226

25. O. Özkan, A. A. Alola, and T. S. Adebayo, 'Environmental benefits of nonrenewable energy efficiency and renewable energy intensity in the USA and EU: examining the role of clean technologies', *Sustain. Energy Technol. Assess.*, vol. 58, p. 103315, Aug. 2023. https://doi.org/10.1016/j.seta.2023.103315
26. J. Wen, D. Zhao, and C. Zhang, 'An overview of electricity powered vehicles: lithium-ion battery energy storage density and energy conversion efficiency', *Renew. Energy*, vol. 162, pp. 1629–1648, Dec. 2020. https://doi.org/10.1016/j.renene.2020.09.055
27. Y. Miao, P. Hynan, A. Von Jouanne, and A. Yokochi, 'Current li-ion battery technologies in electric vehicles and opportunities for advancements', *Energies*, vol. 12, no. 6, p. 1074, Mar. 2019. https://doi.org/10.3390/en12061074
28. X. Shen, H. Liu, X.-B. Cheng, C. Yan, and J.-Q. Huang, 'Beyond lithium ion batteries: higher energy density battery systems based on lithium metal anodes', *Energy Storage Mater.*, vol. 12, pp. 161–175, May 2018. https://doi.org/10.1016/j.ensm.2017.12.002
29. P. Singh, K. Arora, U. C Rathore, G. P. Joshi, and W. Cho, 'Comparative study of controllers in battery energy storage system integrated with doubly fed induction generator-based wind energy conversion system for power quality improvement', *Energy Rep.*, vol. 11, no. 9, pp. 4587–4600, 2024. https://doi.org/10.1016/j.egyr.2024.04.020
30. G. Jeong, H. Kim, J. H. Park, J. Jeon, X. Jin, J. Song, B. R. Kim, M. S, Park, J. M. Kim, and Y. J. Kim, 'Nanotechnology enabled rechargeable Li–SO_2 batteries: another approach towards post-lithium-ion battery systems', *Energy Environ. Sci.*, vol. 8, no. 11, pp. 3173–3180, 2015. https://doi.org/10.1039/C5EE01659B
31. J. Xu, X. Cai, S. Cai, Y. Shao, C. Hu, S. Lu, and S. Ding, 'High-energy lithium-ion batteries: recent progress and a promising future in applications', *Energy Environ. Mater.*, vol. 6, no. 5, p. e12450, Sep. 2023. https://doi.org/10.1002/eem2.12450
32. X. Lai, Q. Chen, X. Tang, Y. Zhou, F. Gao, Y. Guo, R. Bhagat, and Y., 'Critical review of life cycle assessment of lithium-ion batteries for electric vehicles: a lifespan perspective', *eTransportation*, vol. 12, p. 100169, May 2022. https://doi.org/10.1016/j.etran.2022.100169
33. Y. Chen, T. Wang, H. Tian, D. Su, Q. Zhang, and G. Wang, 'Advances in Lithium–Sulfur batteries: from academic research to commercial viability', *Adv. Mater.*, vol. 33, no. 29, p. 2003666, Jul. 2021. https://doi.org/10.1002/adma.202003666
34. M. Rana, N. Alghamdi, X. Peng, Y. Huang, B. Wang, L, Wang, I. R. Gentle, S, Hickey, and B. Luo, 'Scientific issues of zinc-bromine flow batteries and mitigation strategies', *Exploration*, vol. 3, no. 6, p. 20220073, Dec. 2023. https://doi.org/10.1002/EXP.20220073
35. M. Wu, T. Zhao, R. Zhang, H. Jiang, and L. Wei, 'A zinc–bromine flow battery with improved design of cell structure and electrodes', *Energy Technol.*, vol. 6, no. 2, pp. 333–339, Feb. 2018. https://doi.org/10.1002/ente.201700481
36. M. Ulaganathan, S. Suresh, K. Mariyappan, P. Periasamy, and R. Pitchai, 'New zinc–vanadium (Zn–V) hybrid redox flow battery: high-voltage and energy-efficient advanced energy storage system', *ACS Sustain. Chem. Eng.*, vol. 7, no. 6, pp. 6053–6060, Mar. 2019. https://doi.org/10.1021/acssuschemeng.8b06194

37. P. Alotto, M. Guarnieri, and F. Moro, 'Redox flow batteries for the storage of renewable energy: a review', *Renew. Sustain. Energy Rev.*, vol. 29, pp. 325–335, Jan. 2014. https://doi.org/10.1016/j.rser.2013.08.001
38. R. Ye, D. Henkensmeier, S. J. Yoon, Z. Huang, D. K. Kim, Z. Chang, S. Kim, and R. Chen, 'Redox flow batteries for energy storage: a technology review', *J. Electrochem. Energy Convers. Storage*, vol. 15, no. 1, p. 010801, Feb. 2018. https://doi.org/10.1115/1.4037248
39. H. Zhang, W. Lu, and X. Li, 'Progress and perspectives of flow battery technologies', *Electrochem. Energy Rev.*, vol. 2, no. 3, pp. 492–506, Sep. 2019. https://doi.org/10.1007/s41918-019-00047-1
40. H. He, S. Tian, B. Tarroja, O. A. Ogunseitan, S. Samuelsen, and J. M. Schoenung, 'Flow battery production: materials selection and environmental impact', *J. Clean. Prod.*, vol. 269, p. 121740, Oct. 2020. https://doi.org/10.1016/j.jclepro.2020.121740
41. C. M. Julien, 'Advanced materials for electrochemical energy storage: lithium-ion, lithium-sulfur, lithium-air and sodium batteries', *Int. J. Mol. Sci.*, vol. 24, no. 3, p. 3026, Feb. 2023. https://doi.org/10.3390/ijms24033026
42. D. Lei, K. Shi, H. Ye, Z. Wan, Y. Wang, L. Shen, B. Li, Q. H. Yang, F. Kang and Y. B. He, 'Progress and perspective of solid-state lithium–sulfur batteries', *Adv. Funct. Mater.*, vol. 28, no. 38, p. 1707570, Sep. 2018. https://doi.org/10.1002/adfm.201707570
43. A. N. Singh, M. Islam, A. Meena, M. Faizan, D. Han, C. Bathula, A. Hajibabaei, R. Anand, and K. W. Nam, 'Unleashing the potential of sodium-ion batteries: current state and future directions for sustainable energy storage', *Adv. Funct. Mater.*, vol. 33, no. 46, p. 2304617, Nov. 2023. https://doi.org/10.1002/adfm.202304617
44. C. Xian, Q. Wang, Y. Xia, F. Cao, S. Shen, Y. Zhang, M. Chen, Y. Zhong, J. Zhang, X. He, and X. Xia, 'Solid-state electrolytes in lithium–sulfur batteries: latest progresses and prospects', *Small*, vol. 19, no. 24, p. 2208164, Jun. 2023. https://doi.org/10.1002/smll.202208164
45. W. Ni, 'Perspectives on advanced lithium–sulfur batteries for electric vehicles and grid-scale energy storage', *Nanomaterials*, vol. 14, no. 12, p. 990, Jun. 2024. https://doi.org/10.3390/nano14120990
46. F. Zhang, Z. Tang, T. Zhang, H. Xiao, H. Zhuang, X. Liang, L. Zheng, and Q. Gao, 'Enhancing sulfur redox conversion of active iron sites by modulation of electronic density for advanced lithium-sulfur battery', *Small Methods*, vol. 7, no. 10, p. 2300519, Oct. 2023. https://doi.org/10.1002/smtd.202300519
47. J. Sun, T. Wang, Y. Gao, Z. Pan, R. Hu, and J. Wang, 'Will lithium-sulfur batteries be the next beyond-lithium ion batteries and even much better?', *InfoMat*, vol. 4, no. 9, p. e12359, Sep. 2022. https://doi.org/10.1002/inf2.12359
48. L. Huang, J. Li, B. Liu, Y. Li, S. Shen, S. Deng, C. Lu, W. Zhang, Y. Xia, G. Pan, and X. Wang, 'Electrode design for lithium–sulfur batteries: problems and solutions', *Adv. Funct. Mater.*, vol. 30, no. 22, p. 1910375, May 2020. https://doi.org/10.1002/adfm.201910375
49. G. Z. Chen, 'Supercapacitor and supercapattery as emerging electrochemical energy stores', *Int. Mater. Rev.*, vol. 62, no. 4, pp. 173–202, May 2017. https://doi.org/10.1080/09506608.2016.1240914
50. P. Sharma, and V. Kumar, 'Current technology of supercapacitors: a review', *J. Electron. Mater.*, vol. 49, no. 6, pp. 3520–3532, Jun. 2020. https://doi.org/10.1007/s11664-020-07992-4

51. K. Arora, G. P. Joshi, M. Ragab, M. Rawa, A. H. Milyani, R. F. Mansour, and E. Yang, "Bilateral contract for load frequency and renewable energy sources using advanced controller", *Comp. Mat. Cont.*, vol. 73, no. 2, pp. 3165–3180, , 2022. https://doi.org/10.32604/cmc.2022.026966
52. P. Hung, H. Zhang, H. Lin, Q. Guo, K. Lau, and B. Jia, 'Specializing liquid electrolytes and carbon-based materials in EDLCs for low-temperature applications', *J. Energy Chem.*, vol. 68, pp. 580–602, May 2022. https://doi.org/10.1016/j.jechem.2021.12.012
53. A. Ehsani, A. A. Heidari, and H. M. Shiri, 'Electrochemical pseudocapacitors based on ternary nanocomposite of conductive polymer/graphene/metal oxide: an introduction and review to it in recent studies', *Chem. Rec.*, vol. 19, no. 5, pp. 908–926, May 2019. https://doi.org/10.1002/tcr.201800112
54. S. Grover, P. Kadyan, S. Sharma, K. Sharma, and R. K. Sharma, 'Synthesis and characterization of polyaniline nanotube supported nanocomposite of RuO_2 as electrode material for application in supercapacitor device', *Materialia*, vol. 28, p. 101732, May 2023. https://doi.org/10.1016/j.mtla.2023.101732
55. J. Ran, Y. Liu, H. Feng, H. Shi, and Q. Ma, 'A review on graphene-based electrode materials for supercapacitor', *J. Ind. Eng. Chem.*, vol. 137, pp. 106–121, Sep. 2024. https://doi.org/10.1016/j.jiec.2024.03.043
56. S. M. Benoy, M. Pandey, D. Bhattacharjya, and B. K. Saikia, 'Recent trends in supercapacitor-battery hybrid energy storage devices based on carbon materials', *J. Energy Storage*, vol. 52, p. 104938, Aug. 2022. https://doi.org/10.1016/j.est.2022.104938
57. J. Zhang, M. Gu, and X. Chen, 'Supercapacitors for renewable energy applications: a review', *Micro Nano Eng.*, vol. 21, p. 100229, Dec. 2023. https://doi.org/10.1016/j.mne.2023.100229
58. J. Tan, Q. Wu, and X. Zhang, 'Optimal planning of integrated electricity and heat system considering seasonal and short-term thermal energy storage', *IEEE Trans. Smart Grid*, vol. 14, no. 4, pp. 2697–2708, Jul. 2023. https://doi.org/10.1109/TSG.2022.3230016
59. S. Chavan, R. Rudrapati, and S. Manickam, 'A comprehensive review on current advances of thermal energy storage and its applications', *Alex. Eng. J.*, vol. 61, no. 7, pp. 5455–5463, Jul. 2022. https://doi.org/10.1016/j.aej.2021.11.003
60. I. Sarbu, and C. Sebarchievici, 'A comprehensive review of thermal energy storage', *Sustainability*, vol. 10, no. 1, p. 191, Jan. 2018. https://doi.org/10.3390/su10010191
61. B. Akhmetov, A. G. Georgiev, A. Kaltayev, A. A. Dzhomartov, and R. Popov, 'Thermal energy storage systems – review', *Bulg. Chem. Commun.*, vol. 48, pp. 31–40, 2016.
62. S. Sahoo, and P. Timmann, 'Energy storage technologies for modern power systems: a detailed analysis of functionalities, potentials, and impacts', *IEEE Access*, vol. 11, pp. 49689–49729, 2023. https://doi.org/10.1109/ACCESS.2023.3274504
63. E. Doroudchi, H. Khajeh, and H. Laaksonen, 'Increasing self-sufficiency of energy community by common thermal energy storage', *IEEE Access*, vol. 10, pp. 85106–85113, 2022. https://doi.org/10.1109/ACCESS.2022.3195242
64. D. J. Morrison, and S. I. Abdel-Khalik, 'Effects of phase-change energy storage on the performance of air-based and liquid-based solar heating systems', *Sol. Energy*, vol. 20, no. 1, pp. 57–67, 1978. https://doi.org/10.1016/0038-092X(78)90141-X

65. C. Zhao, J. Yan, X. Tian, X. Xue, and Y. Zhao, 'Progress in thermal energy storage technologies for achieving carbon neutrality', *Carbon Neutrality*, vol. 2, no. 1, p. 10, Apr. 2023. https://doi.org/10.1007/s43979-023-00050-y
66. K. Arora, A. Kumar, V. K. Kamboj, D. Prashar, B. Shrestha, and G. P. Joshi, 'Impact of renewable energy sources into multi area multi-source load frequency control of interrelated power system', *MDPI J. Math.*,, vol. 9, no. 2, p. 186, 2021. https://doi.org/10.3390/math9020186
67. J. D. Hunt, B. Zakeri, A. Nascimento, and R. Brandão, 'Pumped hydro storage (PHS)', in *Storing Energy*, Elsevier, 2022, pp. 37–65. https://doi.org/10.1016/B978-0-12-824510-1.00008-8
68. M. T. Ameen, Z. Ma, A. Smallbone, R. Norman, and A. P. Roskilly, 'Demonstration system of pumped heat energy storage (PHES) and its round-trip efficiency', *Appl. Energy*, vol. 333, p. 120580, Mar. 2023. https://doi.org/10.1016/j.apenergy.2022.120580
69. A. Boroomandnia, B. Rismanchi, and W. Wu, 'A review of micro hydro systems in urban areas: opportunities and challenges', *Renew. Sustain. Energy Rev.*, vol. 169, p. 112866, Nov. 2022. https://doi.org/10.1016/j.rser.2022.112866
70. X. Li, and A. Palazzolo, 'A review of flywheel energy storage systems: state of the art and opportunities', *J. Energy Storage*, vol. 46, p. 103576, Feb. 2022. https://doi.org/10.1016/j.est.2021.103576
71. S. Choudhury, 'Flywheel energy storage systems: a critical review on technologies, applications, and future prospects', *Int. Trans. Electr. Energy Syst.*, vol. 31, no. 9, Sep. 2021. https://doi.org/10.1002/2050-7038.13024
72. W. Ji, F. Hong, Y. Zhao, L. Liang, H. Du, J. Hao, F. Fang, and J. Liu, 'Applications of flywheel energy storage system on load frequency regulation combined with various power generations: a review', *Renew. Energy*, vol. 223, p. 119975, Mar. 2024. https://doi.org/10.1016/j.renene.2024.119975
73. Q. Hassan, A. Z. Sameen, H. M. Salman, M. Jaszczur, and A. K. Al-Jiboory, 'Hydrogen energy future: advancements in storage technologies and implications for sustainability', *J. Energy Storage*, vol. 72, p. 108404, Nov. 2023. https://doi.org/10.1016/j.est.2023.108404
74. S. Orlova, N. Mezeckis, and V. P. K. Vasudev, 'Compression of hydrogen gas for energy storage: a review', *Latv. J. Phys. Tech. Sci.*, vol. 60, no. 2, pp. 4–16, Apr. 2023. https://doi.org/10.2478/lpts-2023-0007
75. M. P. Alves, W. Gul, C. A. Cimini Junior, and S. K. Ha, 'A review on industrial perspectives and challenges on material, manufacturing, design and development of compressed hydrogen storage tanks for the transportation sector', *Energies*, vol. 15, no. 14, p. 5152, Jul. 2022. https://doi.org/10.3390/en15145152
76. Y. Qiu, H. Yang, L. Tong, and L. Wang, 'Research progress of cryogenic materials for storage and transportation of liquid hydrogen', *Metals*, vol. 11, no. 7, p. 1101, Jul. 2021. https://doi.org/10.3390/met11071101
77. J. Zheng, H. Zhou, C. G. Wang, E. Ye, J. W. Xu, X. J. Loh, and Z. Li, 'Current research progress and perspectives on liquid hydrogen rich molecules in sustainable hydrogen storage', *Energy Storage Mater.*, vol. 35, pp. 695–722, Mar. 2021. https://doi.org/10.1016/j.ensm.2020.12.007
78. A. Kumar, P. Muthukumar, P. Sharma, and E. A. Kumar, 'Absorption based solid state hydrogen storage system: a review', *Sustain. Energy Technol.* Assess., vol. 52, p. 102204, Aug. 2022. https://doi.org/10.1016/j.seta.2022.102204

79. X. Ding, R. Chen, J. Zhang, W. Cao, Y. Su, and J. Guo, 'Recent progress on enhancing the hydrogen storage properties of Mg-based materials via fabricating nanostructures: a critical review', *J. Alloys Compd.*, vol. 897, p. 163137, Mar. 2022. https://doi.org/10.1016/j.jallcom.2021.163137
80. M. M. Tellez-Cruz, J. Escorihuela, O. Solorza-Feria, and V. Compañ, 'Proton exchange membrane fuel cells (PEMFCs): advances and challenges', *Polymers*, vol. 13, no. 18, p. 3064, Sep. 2021. https://doi.org/10.3390/polym13183064
81. D. K. Madheswaran, M. Thangamuthu, S. Gnanasekaran, S. Gopi, T. Ayyasamy, and S. S. Pardeshi, 'Powering the future: progress and hurdles in developing proton exchange membrane fuel cell components to achieve department of energy goals – a systematic review', *Sustainability*, vol. 15, no. 22, p. 15923, Nov. 2023. https://doi.org/10.3390/su152215923
82. A. Veichtlbauer, C. Praschl, L. Gaisberger, G. Steinmaurer, and T. I. Strasser, 'Toward an effective community energy management by using a cluster storage', *IEEE Access*, vol. 10, pp. 112286–112306, 2022. https://doi.org/10.1109/ACCESS.2022.3216298
83. A. Alzahrani, M. U. Rahman, G. Hafeez, G. Rukh, S. Ali, S. Murawwat, F. Iftikhar, S. I. Haider, M. I. Khan, and A. M. Abed, 'A strategy for multi-objective energy optimization in smart grid considering renewable energy and batteries energy storage system', *IEEE Access*, vol. 11, pp. 33872–33886, 2023. https://doi.org/10.1109/ACCESS.2023.3263264
84. M. Rouholamini, C. Wang, H. Nehrir, X. Hu, Z. Hu, H. Aki, B. Zhao, Z. Miao, and K. Strunz, 'A review of modeling, management, and applications of grid-connected li-ion battery storage systems', *IEEE Trans. Smart Grid*, vol. 13, no. 6, pp. 4505–4524, Nov. 2022. https://doi.org/10.1109/TSG.2022.3188598
85. H. Zandi, M. Starke, C. Winstead, T. Kuruganti, J. Hill, and F. Li, 'Optimal operation of integrated pv and energy storage considering multiple operational modes with a real-world case study', *IEEE Access*, vol. 11, pp. 99070–99082, 2023. https://doi.org/10.1109/ACCESS.2023.3313502
86. K. Arora, A. Kumar, V. K. Kamboj, D. Prashar, S. Jha, B. Shrestha, and G. P. Joshi, 'Optimization methodologies and testing on standard benchmark functions of load frequency control for interconnected multi area power system in smart grids', *MDPI J. Math.*, , vol. 8, no. 6, p. 980, 2020. https://doi.org/10.3390/math8060980.
87. S. Karrari, N. Ludwig, G. De Carne, and M. Noe, 'Sizing of hybrid energy storage systems using recurring daily patterns', *IEEE Trans. Smart Grid*, vol. 13, no. 4, pp. 3290–3300, Jul. 2022. https://doi.org/10.1109/TSG.2022.3156860
88. M. K. Khan Prince, M. T. Arif, A. Gargoom, M. W. Altaf, K. M. Muttaqi, and M. E. Haque, 'Coordinated control of grid-connected pmsg based wind energy system with STATCOM and supercapacitor energy storage', in 2022 IEEE Industry Applications Society Annual Meeting (IAS), Detroit, MI, USA: IEEE, Oct. 2022, pp. 01–08. https://doi.org/10.1109/IAS54023.2022.9939705

Chapter 8

Conventional energy sources in wireless communication

A challenge

Shakti Raj Chopra

8.1 INTRODUCTION

In the context of wireless communication, conventional energy sources primarily refer to those utilized to power wireless network infrastructure and devices. The main energy sources in these systems are electricity delivered by traditional power systems and non-renewable fossil fuels.

In conventional systems, the primary energy sources include coal, oil, and natural gas. The detrimental environmental implications of these energy sources, such as greenhouse gas emissions and global warming, contribute to environmental degradation. Due to the rapid growth of wireless communication, there will inevitably be a link between network expansion and environmental harm unless alternative energy sources are explored [1].

Significant amounts of energy are required by a conventional cellular base station to convey messages, maintain connectivity, and manage growing traffic demands. Many network providers are adopting renewable energy sources, such as solar and wind, to power wireless infrastructure in response to rising energy costs and environmental concerns. These renewable energy sources help reduce the carbon footprint of wireless communication networks [2].

Key elements of wireless communication that use traditional energy sources include the following:

- Base Stations: They consume substantial energy to guarantee uninterrupted network connectivity.
- Data Centers: At the core of cloud services, data centers require a lot of electricity to operate cooling systems and computers.
- Network Infrastructure: Energy-intensive antenna towers and transmission equipment are used to ensure that signals can travel great distances.

In response to these difficulties, interest in sustainable and renewable energy sources—such as solar, wind, and hydrogen fuel cells—for enabling wireless communications has grown. These sources aim to improve energy efficiency and lower carbon emissions [3]. In conclusion, the transition

DOI: 10.1201/9781003542568-8

toward more sustainable energy options is gaining momentum to address environmental and economic problems, even though conventional energy sources have provided the groundwork for the current wireless communication ecosystem.

8.2 OVERVIEW OF CONVENTIONAL ENERGY SOURCES IN WIRELESS COMMUNICATION

Conventional energy sources have provided the foundation for the power supply of wireless communication networks. Among them are

- Fossil fuels, such as coal, oil, and natural gas, are utilized to produce electricity for wireless infrastructure, which includes network operation centers, base stations, and data centers.
- Diesel generators are frequently used as backup power sources, particularly in remote locations where grid electricity is unstable or nonexistent.
- Nuclear Energy: Although it is less frequent in telecommunications, some areas power wireless systems using electricity generated from nuclear sources.

Although these energy sources have made it easier for wireless infrastructure to grow quickly, they also have a number of drawbacks.

- High Carbon Emissions: The use of fossil fuels significantly contributes to greenhouse gas emissions.
- Environmental Degradation: The mining and use of coal, oil, and gas negatively impacts ecosystems and biodiversity.
- Increasing Costs: Variations in fuel prices can lead to higher operating expenses for telecom firms.

8.3 CHALLENGES AND POSSIBLE SOLUTIONS

8.3.1 Challenges

Systems that use wireless communication, particularly those that utilize 5G and beyond, consume a large amount of energy to operate. Traditional energy sources lead to high operating expenses and are not energy-efficient. Base stations, data centers, and network infrastructure, for instance, require substantial power, which is often supplied by non-renewable sources. This inefficiency impacts the overall sustainability of wireless communication networks [4].

8.3.2 Effect on the environment

There is a significant carbon footprint associated with wireless infrastructure that is powered by fossil fuels. Data centers and mobile networks require massive energy consumption, and when that energy comes from conventional sources, it increases greenhouse gas emissions, exacerbates climate change, and degrades the environment.

8.3.3 Availability and dependability of energy

For wireless communication networks to function properly, electricity must be steady and dependable. Conventional energy sources can be erratic or prone to disruptions in many areas. These disruptions can include blackouts brought on by dwindling resources, unstable economies, or geopolitical conflicts. The ongoing operation of vital wireless services, such as emergency communications, is threatened by this unreliability.

8.3.4 Increasing energy prices

The price of energy from conventional sources is predicted to climb in tandem with the world's growing energy demand and the depletion of fossil fuel supplies. As a result, wireless network providers will face financial strain, having to power more base stations, small cells, and other equipment to keep up with the rising demand for data. Rising energy prices could lead to increased operating costs and higher consumer prices.

8.3.5 Expanding the network and scalability

As wireless networks develop with the advent of 5G, Internet of Things (IoT), and soon-to-be 6G technologies, more network infrastructure—such as additional base stations and edge computing nodes—is required. Because of finite resources and increasing environmental effects, scaling these systems using traditional energy sources is not viable.

8.3.6 Battery durability and energy reclamation

Batteries are essential for the operation of mobile devices, sensors, and other end-user equipment in wireless communication. Conventional power grids frequently provide the energy needed to charge billions of devices, increasing the demand for fossil fuel energy sources. In the Internet of Things and sensor networks, where energy efficiency and endurance are crucial, limited battery life presents a significant challenge. These needs are best met by energy harvesting technology or renewable energy sources, which are more difficult to support than traditional sources [5].

8.3.7 The influence of policies and regulations

To achieve climate targets and lessen the environmental impact of industries, governments and international organizations are increasingly advocating for a shift to cleaner, renewable energy sources. Wireless network operators that depend on non-renewable energy sources could face financial penalties, policy limits, or regulatory pressure to reduce carbon emissions.

8.3.8 Energy consumption in developing areas

Reliable traditional energy sources are either hard to come by or extremely expensive in many developing regions. This makes it more difficult to install wireless networks, particularly in isolated or rural locations. Thus, the dependence on traditional energy sources hampers efforts to increase digital connectivity and ensure equitable access to communication technologies.

Wireless communication systems' environmental effect, scalability, and sustainability are severely hampered by conventional energy sources. To establish a more resilient and sustainable wireless network infrastructure, there is an increasing need to move toward energy-efficient technology, renewable energy sources (such as solar and wind), and breakthroughs in green communications [6,7].

8.4 POSSIBLE REMEDIES AND SUBSTITUTES

- Integration of Renewable Energy:
  - Solar Power: By adding solar panels to base stations and data centers, dependence on diesel generators or the grid for electricity can be reduced. In many places, solar energy is abundant and clean.
- Wind Energy: Wind turbines can be installed to provide a renewable energy source for wireless communication infrastructure, especially in rural or coastal areas.
- Hydroelectric Power: Wireless networks can be sustainably powered by hydroelectric energy in areas where it is feasible.
- Hybrid systems, which combine solar, wind, and traditional grid electricity, ensure a steady supply of power even if one of the sources is unavailable.

8.5 METHODS OF ENERGY HARVESTING

- Radio Frequency Energy Harvesting: In wireless sensor networks, radio frequency signals from ambient sources, such as TV towers, Wi-Fi routers, and cellular base stations, can be collected and converted into useful energy for low-power devices.

- Vibration and Kinetic Energy Harvesting: In remote or difficult-to-reach locations, devices can harness mechanical vibrations or human motion to generate energy for wireless communication equipment.
- Thermal Energy Harvesting: Variations in temperature can be exploited to produce energy to power wireless sensors or small devices, especially in outdoor settings.

8.6 ENERGY-EFFICIENT NETWORK DESIGN

- Densification of the network: By deploying smaller, more energy-efficient base stations—such as microcells, picocells, and femtocells—than conventional large towers, energy consumption can be reduced. Additionally, by improving coverage and data throughput, these smaller cells can enhance network efficiency.
- Sleep Mode Technologies: When traffic demand is low, base stations and other network components can be configured to enter low-power or sleep modes, significantly reducing energy usage during off-peak hours.
- Cognitive Radio Networks: Cognitive radio networks enable devices to automatically adjust their communication to available spectrum resources, which can lead to increased energy efficiency through dynamic spectrum access.
- Thermal Energy Harvesting: Variations in temperature can be used to produce energy to power wireless sensors or tiny equipment, especially in outdoor settings.

8.7 SOPHISTICATED PROTOCOLS OF COMMUNICATION

- Energy-Aware Routing: By selecting data transmission channels that are energy-efficient, energy-aware routing protocols in wireless sensor networks aim to reduce energy consumption.
- Optimized Medium Access Control Protocols: Optimized medium access control protocols can decrease energy-wasting packet collisions and idle listening times.
- Green Network Protocols: These protocols apply techniques such as load balancing, adaptive modulation, and dynamic power adjustment to lower energy consumption across various network layers, from the physical layer to the application layer [8].

8.8 INCREASING HARDWARE PERFORMANCE

- Energy-Efficient Amplifiers: Power amplifiers used in wireless communication equipment can function more effectively by reducing power dissipation and increasing conversion efficiency.
- Advanced Semiconductor Technologies: The energy efficiency of transceivers can be enhanced by using low-power semiconductor materials (such as GaN or SiC), which decreases energy waste and improves system performance.
- Overall Battery and Power Management: Techniques like dynamic power adjustment, adaptive modulation, and load balancing can reduce energy usage across different layers of the network, from the physical layer to the application layer. In addition, efficient power management systems for wireless devices, including advanced battery technologies and energy-efficient chargers, can contribute to these reductions [9].

8.9 EDGE AND FOG COMPUTING

- Offloading to the Edge: Processing data at the network's edge, where it is closer to end users, reduces the amount of data that needs to be sent back to central servers. Consequently, this leads to lower energy consumption during data transmission over long distances.
- Fog Computing: Similar to edge computing, this approach distributes networking, storage, and processing resources throughout the network. By enabling local data processing and decreasing dependence on energy-intensive data centers, it helps lower overall energy consumption.

8.10 USING AI AND MDL TO IMPROVE ENERGY EFFICIENCY

- Predictive analytics: Employing machine learning algorithms to forecast network traffic patterns allows for real-time adjustments in power consumption to enhance energy efficiency.
- Energy-Efficient Resource Allocation: AI can minimize wasteful energy usage by dynamically allocating resources (such as electricity and bandwidth) based on user demand, energy availability, and current network conditions [10].

8.11 CONCLUSION

There are several problems with wireless communication's continuous reliance on traditional energy sources, such as increased operational costs, inefficiency, and environmental damage. The industry must investigate sustainable energy options as the demand for wireless connectivity rises due to the emergence of 5G, IoT, and other technologies. The telecom sector faces energy challenges that must be addressed. These issues can be met by implementing smart energy management systems, improving energy-efficient network designs, and integrating renewable energy sources. To ensure the long-term sustainability of wireless communication systems, a holistic approach is necessary, combining technological innovation, policy reform, and a commitment to reducing the environmental impact of energy consumption. Integrating renewable energy with energy-efficient network architecture and cutting-edge technology is essential for mitigating the environmental impact of traditional energy sources in wireless communication. To address the increasing demand for connectivity without depleting natural resources, these technologies can be used to construct more affordable and sustainable wireless communication systems.

REFERENCES

1. Andrews, J. G., Buzzi, S., Choi, W., Hanly, S. V., Lozano, A., Soong, A. C., and Zhang, J. C. "What will 5G be?" *IEEE Journal on Selected Areas in Communications*, vol. 32, no. 6, pp. 1065–1082, June 2014.
2. Shahriar, M., and Chopra, S. R. "Solar panel based wireless power bank works in a small satellite power management system." In *Smart Small Satellites: Design, Modelling and Development: Proceedings of the International Conference on Small Satellites, ICSS 2022*, pp. 147–153. Singapore: Springer Nature Singapore, 2023.
3. Rahul, S., Chopra, R. R., and Gupta, A. "UAV assisted next generation wireless communication network." In *2024 International Conference on Emerging Smart Computing and Informatics (ESCI)*, pp. 1–5. Pune, India: IEEE, 2024. https://doi.org/10.1109/ESCI59607.2024.10497324
4. Singh, P., Arora, K., Rathore, U. C, Joshi, G. P. and Cho, W. "Comparative study of controllers in battery energy storage system integrated with doubly fed induction generator-based wind energy conversion system for power quality improvement", *Energy Reports*, vol. 11, no. 9, pp. 4587–4600, 2024. https://doi.org/10.1016/j.egyr.2024.04.020
5. Sai, A. R. V., Sunith, Y. E., Sai, A. R. N., Pranav, M., and Chopra, S. R. "Design and surveillance system using wireless technologies check for updates." ICT Infrastructure and Computing: Proceedings of ICT4SD 2023, vol. 3, no. 754, p. 305, 2023.
6. Jiang, W., and Hanzo, L. "Artificial intelligence-powered next-generation wireless networks." *IEEE Wireless Communications*, vol. 26, no. 5, pp. 10–16, 2017.
7. Zhang, H., Liu, N., Chu, X., Long, K., Aghvami, A. H., and Leung, V. C. M. "Network slicing based 5G and future mobile networks: mobility, resource management, and challenges." *IEEE Communications Magazine*, vol. 55, no. 8, pp. 138–145, 2017.

8. Sharma, R., Chopra, S. R., Gupta, A., Kaur. R., Tanwar, S., Pau, G., Sharma, G., Alqahtani, F., and Tolba, A. "Deployment of unmanned aerial vehicles in next-generation wireless communication network using multi-agent reinforcement learning," *IEEE Access*, vol. 1, pp. 1–16, 2024.
9. Kaloxylos, A., Alonistioti, N., and Burke, M. Leveraging AI in wireless networks: present and future trends. *IEEE Network*, vol. 35, no. 2, pp. 10–17, 2021.
10. Sharma, A., Chopra, S. R., Sapate, S. G., Arora, K., Rahmani, M. K. I., Jha, S., Ahmad, S., Ahmed, Md E., Abdeljaber, H. A J., and Nazeer, J. "Artificial intelligence techniques for landslides prediction using satellite imagery," *IEEE Access*, vol. 1, pp. 1–23, 2024.

Chapter 9

Non-conventional energy resources and the role of mechatronics systems design in enhancing manufacturing resilience

Praveen Kumar Malik

9.1 INTRODUCTION TO MECHATRONICS

Mechatronics is an interdisciplinary field that combines mechanical engineering, electrical engineering, and computer engineering. It deals with the design, development, implementation, and maintenance of automated mechanical systems. Mechatronics is used to create robots, automated manufacturing systems, medical equipment, and a wide range of other advanced technologies. The main objective of mechatronics is to produce systems that are reliable, efficient, and cost-effective. The components of mechatronics include sensors, actuators, controllers, and software. Sensors are used to detect changes in the environment, while actuators are used to cause physical actions. Controllers process the data from the sensors and control the actuators. Finally, software is used to program the system.

9.2 FUNDAMENTALS OF MECHATRONICS

Mechatronics is the combination of mechanical engineering, electronics, and computer control systems. This combination gives rise to products that can sense, respond, and interact with their environment. Mechatronics integrates these three disciplines to create automated systems, such as robots and automated machinery, that can make decisions and carry out complex tasks. In the modern world, mechatronics is utilized in a variety of industries, from automotive manufacturing to consumer electronics. This technology has revolutionized the way we work and live. The fundamentals of mechatronics include the principles of mechanical engineering, electronics, and computer control systems, as well as their application in modern engineering. Topics may include system design, operation, and troubleshooting, alongside the fundamentals of sensors, actuators, and power transmission. Additionally, modern mechatronic systems require an understanding of digital control systems and computer programming.

DOI: 10.1201/9781003542568-9

9.3 MECHATRONIC SYSTEMS DESIGN

Mechatronics and digital control systems are employed to control and automate complex machines. They utilize a combination of mechanical, electrical, and software engineering to create sophisticated and efficient processes. These systems are found in various industries, including automotive, aerospace, and medical. They are designed to respond quickly and accurately to changes in the environment, such as temperature or pressure. They also monitor and adjust the performance of various components in a machine. These systems can control motors, actuators, sensors, and other hardware components. They can also monitor and manage the performance of software applications. The performance of mechatronics can be monitored and controlled using a variety of methods, depending on the specific system. For example, sensors can measure variables such as temperature, pressure, speed, and position, while feedback can adjust the system accordingly. Automated control systems can also adjust system parameters in response to changes in the environment or desired performance. Additionally, software programs can monitor and analyze system performance, allowing users to identify areas for improvement and optimize the system for maximum efficiency.

9.3.1 How to optimize the system efficiency of mechatronics

1. Improve System Design: Utilize modern design tools and techniques to create more efficient systems with fewer components and lower power requirements.
2. Improve System Components: Incorporate more energy-efficient components such as motors, actuators, sensors, and other subsystems.
3. Implement Energy Management Strategies: Implement energy management strategies such as variable speed drives and predictive maintenance to lower energy consumption.
4. Utilize Automation: Utilize automation to reduce manual processes, reduce labor costs, and improve system efficiency.
5. Implement Real-Time Monitoring: Establish real-time monitoring of system components and performance metrics to identify potential areas for improvement.
6. Utilize AI and Machine Learning: Apply AI and machine learning to automatically detect and optimize system performance.

9.4 ROBOTICS IN MECHATRONICS

Robotics in mechatronics refers to the application of robotic technology in the field of mechatronics, which integrates mechanical, electrical, and computer engineering. It combines robotics, computer engineering, and other disciplines such as materials engineering, software engineering, and manufacturing engineering to create robots capable of performing complex tasks. Robots in mechatronics are used across various industries, from manufacturing to medical research. They automate processes, reduce labor costs, and increase productivity. Robotics in mechatronics also encompasses artificial intelligence, which enables robots to respond to their environment and make decisions. Some common applications include industrial automation, factory automation, and medical robotics. Industrial and factory automation utilize mechatronics to control and streamline processes within a factory or industrial setting. Mechatronics involves integrating mechanical systems, electronic control systems, and software engineering. This interdisciplinary field focuses on the synergy of mechanical and electrical components and systems, which are employed to control and monitor machines and processes in industrial environments. This includes computer numerical control (CNC) systems, programmable logic controllers, robotics, drives and motors, sensors and actuators, and system integration. Mechatronics is also used to design custom automation solutions for specific applications, helping to increase efficiency, reduce cycle times, improve quality, and lower costs.

9.5 CONTROL SYSTEMS FOR MECHATRONICS

Control systems for mechatronics are used to manage and control the behavior of a mechatronic system. These systems typically include components such as sensors, actuators, and controllers, and are utilized to regulate the motion and operation of a mechatronic system. The use of control systems in mechatronics allows for precise, consistent performance along with improved safety and reliability. Common control systems used in mechatronic systems include proportional, integral, and derivative (PID) controllers, fuzzy logic controllers, and neural network controllers. Each of these control systems has its own benefits and drawbacks, depending on the specific application. Fuzzy logic controllers, neural network controllers, and PID controllers are all used in mechatronics to help control the movement of a system in order to achieve a desired output. Fuzzy logic controllers use fuzzy logic and inference rules to manage a system. Neural network controllers use artificial intelligence to interpret data and adjust a system accordingly. PID controllers apply the PID algorithm to modify a system's parameters. All three of these controllers are used in mechatronics to help create the desired output.

9.6 SENSORS AND ACTUATORS IN MECHATRONICS

Sensors and actuators are essential components of mechatronics. Sensors are used to measure and detect physical quantities such as temperature, pressure, force, position, and velocity. Actuators convert energy into motion, such as motors and solenoids. Together, sensors and actuators enable mechatronic systems to sense the environment, process data, and react accordingly. For example, in a robotic arm, sensors measure the position of the arm, and actuators move the arm to its desired position. In an automated factory, sensors monitor the production process, and actuators adjust the speed of the machines.

9.7 ADVANCED MECHATRONIC CONCEPTS

Advanced mechatronic concepts involve the use of advanced technology to create complex, highly integrated, automated systems. These systems are used across various industries to enhance efficiency and safety, as well as to reduce production costs. Mechatronic concepts typically integrate mechanical, electrical, and software components into a single system. Examples of advanced mechatronic concepts include robotics, automated production systems, and CNC machines. CNC machines are automated manufacturing tools that are programmed using numerical control, enabling them to produce precision parts with high accuracy and repeatability. CNC machines can create parts in a fraction of the time it would take to produce them manually, and they are used in several industries, including aerospace, automotive, and medical device manufacturing. Operated by a computer program, CNC machines can perform a variety of tasks, including cutting, drilling, milling, and grinding. They are often used in combination with other automated processes, such as robotic arms and automated assembly lines.

9.8 APPLICATIONS OF MECHATRONICS IN INDUSTRY

The following are some applications of mechatronics in industry:

1. Automated Manufacturing: Mechatronics can be used to automate a wide range of manufacturing processes, from assembly lines to packaging and shipping functions. Automated systems reduce the need for human labor and increase production speed, accuracy, and efficiency.
2. Automated Inspection: Mechatronics can be used to automate the inspection of manufactured goods, ensuring that products meet quality standards. Automated inspection systems can inspect products at high speeds and with greater accuracy than human inspectors.
3. Robotics: Robotics is an application of mechatronics that involves the design, construction, and operation of robots. This field enables machines to move, manipulate, and interact with their environment, allowing for a wide range of automation applications.
4. Autonomous Vehicles: Autonomous vehicles represent another application of mechatronics. These vehicles use sensors, cameras, and GPS

technology to autonomously navigate their environment, enabling a variety of applications, such as delivery services or transportation.

5. Medical Devices: Mechatronics can also be used to develop medical devices, including prosthetics, artificial organs, and surgical robots. These devices provide more accurate and precise treatments, contributing to improved patient outcomes.

9.9 CONCLUSION

In addition, mechatronics systems can be used to monitor and control the production process, allowing manufacturers to identify and address potential problems that may arise. This helps manufacturers detect risks before they become issues and take steps to mitigate them. It can reduce the risk of costly production interruptions and increase the resilience of the manufacturing process. Finally, mechatronics systems can automate certain parts of the production process. This reduces the amount of manual labor needed and decreases the time required to complete tasks, benefiting manufacturers by cutting costs and increasing efficiency.

Overall, mechatronics system design has the potential to significantly enhance manufacturing resilience. By enabling manufacturers to respond quickly to changes in the market, adjust production lines rapidly, and minimize downtime, mechatronics systems can help them remain competitive and resilient in the face of changing market conditions.

FURTHER READINGS

1. K. Kuru and H. Yetgin, "Transformation to advanced mechatronics systems within new industrial revolution: a novel framework in automation of everything (AoE)," *IEEE Access*, vol. 7, pp. 41395–41415, 2019. https://doi.org/10.1109/ACCESS.2019.2907809
2. Y. Feng, M. Zhou, Q. -L. Han, F. Han, Z. Cao and S. Ding, "Integral-type sliding-mode control for a class of mechatronic systems with gain adaptation," *IEEE Transactions on Industrial Informatics*, vol. 16, no. 8, pp. 5357–5368, Aug. 2020. https://doi.org/10.1109/TII.2019.2954550
3. E. Csencsics and G. Schitter, "Exploring the pareto fronts of actuation technologies for high performance mechatronic systems," *IEEE/ASME Transactions on Mechatronics*, vol. 26, no. 2, pp. 1053–1063, April 2021. https://doi.org/10.1109/TMECH.2020.3016087
4. T. Yang, X. Zhang, Y. Fang, N. Sun and M. Iwasaki, "Observer-based adaptive fuzzy event-triggered control for mechatronic systems with inaccurate signal transmission and motion constraints," *IEEE/ASME Transactions on Mechatronics*, vol. 27, no. 6, pp. 5208–5221, Dec. 2022. https://doi.org/10.1109/TMECH.2022.3175969

5. T. Staessens, T. Lefebvre and G. Crevecoeur, "Adaptive control of a mechatronic system using constrained residual reinforcement learning," *IEEE Transactions on Industrial Electronics*, vol. 69, no. 10, pp. 10447–10456, Oct. 2022. https://doi.org/10.1109/TIE.2022.3144565
6. L. Celentano, M. V. Basin and P. Shi, "Majorant-based control methodology for mechatronic and transportation processes," *IEEE Access*, vol. 9, pp. 157916–157939, 2021. https://doi.org/10.1109/ACCESS.2021.3126123
7. P. Singh, K. Arora, U. C Rathore, G. P, Joshi and W, Cho, "Comparative study of controllers in battery energy storage system integrated with doubly fed induction generator-based wind energy conversion system for power quality improvement," *Energy Reports*, 2024, vol. 11, no. 9, pp. 4587–4600, 2024. https://doi.org/10.1016/j.egyr.2024.04.020
8. S. Lin, L. Jia, Y. Wang and H. Zhang, "A new function-topology-based method for assessing passive safety of mechatronics systems," *IEEE Access*, vol. 8, pp. 9312–9324, 2020. https://doi.org/10.1109/ACCESS.2020.2965588
9. A. S. Duggal, P. K. Malik, A. Gehlot, R. Singh, G. S. Gaba, M. Masud and J. F. Al-Amri, "A sequential roadmap to industry 6.0: exploring future manufacturing trends," *IET Communication*, pp. 1–11, 2021. https://doi.org/10.1049/cmu2.12284
10. P. K. Malik, R, Sharma, R. Singh, A. Gehlot, S. C. Satapathy, W. S. Alnumay, D. Pelusi, U. Ghosh and J. Nayak, "Industrial internet of things and its applications in Industry 4.0: state of the art," *Computer Communications*, vol. 166, pp. 125–139, 2021. ISSN 0140–3664, https://doi.org/10.1016/j.comcom.2020.11.016
11. P. V. K. Sneha, N. Bilandi and A. Gupta, "Narrow band-IoT and long-range technology of IoT smart communication: designs and challenges," *Computers & Industrial Engineering*, vol. 172, Part A, p. 108572, 2022. ISSN 0360-8352, https://doi.org/10.1016/j.cie.2022.108572

Chapter 10

Biofuels and biogas

B. Namratha and Santosh L. Gaonkar

10.1 INTRODUCTION TO BIOFUELS

The advancement of biofuels is experiencing a striking surge due to the substantial rise in global oil demand. The development of biofuels from an elementary level has the potential to instigate rural progress, achieve energy autonomy, and mitigate volatile climatic variations occurring worldwide. These aspects also encourage political initiatives aimed at societal welfare and securing energy resources (Gustavo de et al., 2017).

Biodiesel, bioethanol, and biobutanol are quintessential examples of biofuels (agrofuels) that derive from agriculture and forestry. These biofuels play a critical role in reducing global greenhouse gas emissions (Rafael et al., 2008). Agricultural and forestry residues are essential in regulating the carbon dioxide cycle, as they absorb carbon dioxide during their growth and release it upon combustion. Consequently, biofuels indirectly contribute to offsetting carbon dioxide emissions (Indu et al., 2018).

The escalating demand for biofuels in various nations, along with utilization patterns, is directly proportional to production and manufacturing techniques. It is essential to thoroughly assess agricultural resources to ensure the sustainable management of biofuel needs. The surge in consumption is intricately tied to land productivity and the quantity of crops cultivated in specific areas designated for biofuel production (Morshadul et al., 2023).

Beyond agricultural land, factors such as biomass availability, fluctuating oil prices, and governmental subsidies play a fundamental role in determining the future trajectory and expansion rate of global biofuel consumption. Technological advancements should be taken seriously, as they can lower the costs associated with second-generation biofuels (Singh et al., 2024).

Rising commodity prices have increased these expenditures. Feedstock costs represent the primary expense in biofuel production, as evidenced by sugarcane in Brazil, corn in the United States, and sugar beet in the European Union. Energy constitutes another significant cost component, making up a substantial portion of operational expenses in certain nations. To address the concerns of investors and producers regarding

DOI: 10.1201/9781003542568-10

fluctuating feedstock prices, energy costs, and biofuel selling prices, governments have implemented policy instruments to mitigate risks and uncertainties. The most prevalent approach is a mandate requiring the blending of biofuel with fossil fuels, ensuring a stable market for biofuels (Carlo and André, 2006).

Nations utilize subsidies, tax incentives, and import limitations to counterbalance the elevated production costs of biofuels and stimulate their utilization. Biofuels will be integrated into a broader scheme aimed at addressing soaring oil prices, which includes conservation efforts and alternative fuels; yet their global impact will be limited due to land requirements. In the United States, substituting gasoline with ethanol would require a greater land area than is currently used for all agricultural purposes. Technological advancements, particularly in the utilization of energy from cellulose materials, have the potential to substantially enhance biofuel yields and reduce land prerequisites (Michael et al., 2007).

10.2 HISTORY AND DEVELOPMENT OF BIOFUELS

The transesterification of vegetable oils, a procedure with origins dating back to the 19th century, serves as the foundation for the global manufacture of biofuels. Conventional sources such as peanut, hemp, corn oil, and animal tallow are presently being complemented or substituted by soybean, rapeseed, forest residues, trees, and sugar cane (Ayadi et al., 2016).

Historically, liquid biofuels like ethanol and peanut oils were utilized by Nikolaus August Otto and Rudolph Diesel to ignite engines. Henry Ford, the founder of the Ford Corporation, engineered his vehicles to operate on ethanol and advocated fervently for the advancement of sustainable energy sources. Biomass fuels were utilized during the Second World War due to significant oil shortages (Agarwal and Kumar, 2018).

In 1973, the initial occurrence of dual oil crises, orchestrated by the Organisation of the Petroleum Exporting Countries (OPEC)—a dominant entity in the global oil market—deliberately curtailed production levels and escalated prices. The subsequent crisis followed in 1978. These incidents served as a catalyst for the revival of biofuels, prompting a proliferation of bioenergy research, developmental endeavors, and deployment initiatives, complemented by global endeavors aimed at their advancement. As a consequence, the United Nations (UN) International Biofuels Forum materialized, spearheaded by Brazil, China, India, South Africa, the United States, and the European Commission. Moreover, non-governmental educational entities arose to advocate for biofuels through interactive demonstrations, educational presentations, and outreach initiatives. There has been substantial backing from the populace, governmental bodies, industry stakeholders, and segments of the agricultural sector in championing biofuels, with the ultimate objective of transitioning from the fossil fuel-dominated landscape of the preceding five decades to a bio-based sector capable of ensuring a more stable energy supply (Mingxin et al., 2015).

10.3 TYPES OF BIOFUELS

Many researchers have categorized biofuels in different ways. Based on the commercialization status of biofuel technologies, biofuels can be divided into conventional biofuels and advanced biofuels (Figure 10.1). They can be further categorized into commercial, early commercial, demonstration, and research according to their level. This can be depicted in the following figure. Advanced biodiesels are currently not extensively available; however, there is a possibility that they will achieve full commercialization in the near future.

Most scholars have classified biofuels as primary and secondary biofuels. Primary biofuels, known as natural biofuels, are typically sourced

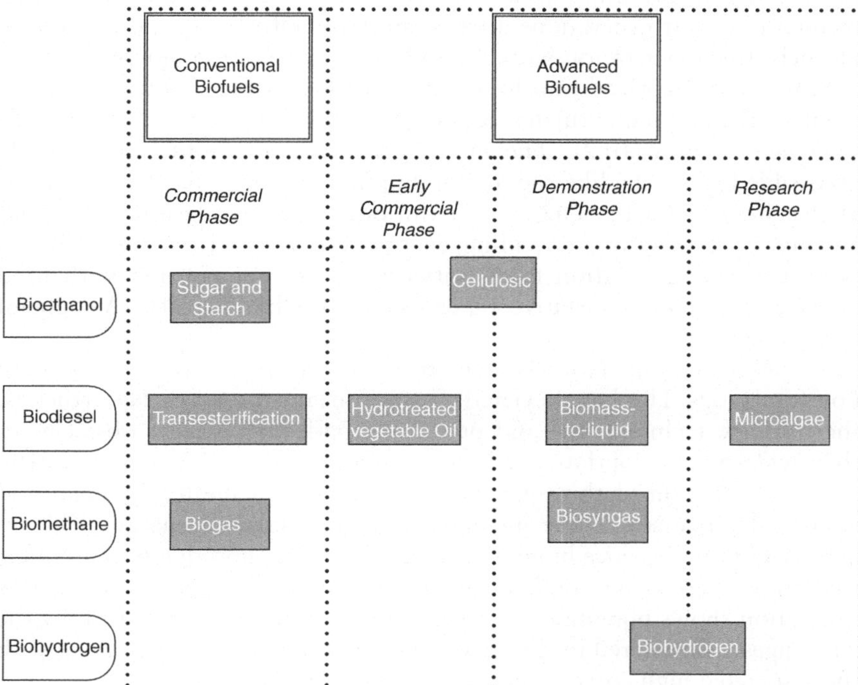

Figure 10.1 Categorization of biofuels based on commercialization.

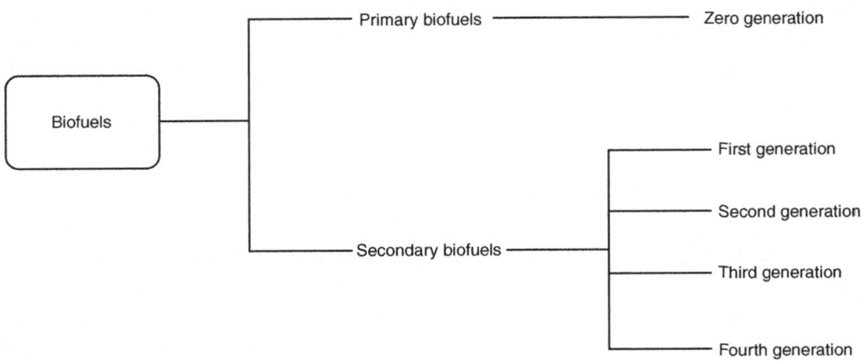

Figure 10.2 Classification of biofuels into different generations.

from organic materials such as firewood, plants, vegetables, animal waste, and landfill gas. These organic biofuels find common applications in cooking, heating, brick kilns, or electricity generation, and are often referred to as zero-generation biofuels. In contrast, secondary biofuels can be segmented into four generations: first-generation biofuels, second-generation biofuels, third-generation biofuels, and fourth-generation biofuels.

Biofuels, categorized into four generations, signify a gradual advancement in their manufacturing, feedstock utilization, and ecological consequences (Figure 10.2). The initial generation of biofuels uses easily accessible feedstocks like sugar, starch, and vegetable oils. These materials, such as soybean, maize, and palm oil, can be easily integrated into current infrastructure, providing energy security and biodegradability. However, this generation faces criticism due to competition with food crops and potential negative repercussions on biodiversity (Awogbemi et al., 2021).

Second-generation biofuels stem from wheat straw, wood, and other non-food crops. This has alleviated pressure on land resources. Nevertheless, the intricate technological and processing pathways present obstacles to their widespread adoption on a large commercial scale (Roger et al., 2019).

The production of third-generation biofuels is facilitated by algal and microbial processes. These methods result in increased energy yield per acre. Additionally, since biomass can be grown on unproductive and infertile lands, there is no conflict with food crop cultivation. Although this generation shows potential, widespread implementation is hindered by the challenges encountered in the harvesting and processing stages, along with the associated high costs (Acheampong et al., 2017).

Green diesel and biogasoline known as fourth-generation biofuels are recognized for their reduced carbon footprint and utilization of synthetic raw materials. However, these biofuels remain in an early developmental phase, necessitating further research endeavors to overcome the high costs and technological obstacles for full utilization (Zul, 2022).

Biofuels can also be classified based on the raw materials utilized, including first-generation, 1.5-generation, second-generation, and third-generation. First-generation biofuels rely on crops like corn and wheat, while 1.5-generation biofuels require oil-bearing crops such as jatropha and non-food crops like cassava. Second-generation biofuels are derived from sources like grass, straw, forest residue, and cooking oil, whereas third-generation biofuels depend on algal feedstock (Sakthivel et al., 2023).

10.4 BIOMASS SOURCES FOR BIOFUELS

A thorough literature survey reveals that a single source may participate in multiple biofuel production processes (Williams et al., 2016).

Feedstocks with high oil content, such as sunflower, soybean, and rapeseed, are categorized as primary. The effective extraction of oil from these sources relies on factors like botanical origin, as well as the tools and methods employed during extraction (Günnur and Nilgün, 2013).

Lignin-rich wood, rice straw, and sugarcane bagasse fall under the second category. The transformation process is hindered by initial pre-treatment stages due to the resistance offered by lignin at this phase.

Municipal solid waste and animal manure, characterized by high moisture levels, constitute the third group of feedstocks. While these materials exhibit favorable outcomes for digestion and extraction techniques, they pose challenges for combustion and gasification processes (Ravi et al., 2021).

Algae, abundant in lipid content, demonstrate rapid growth rates, possess a simple cellular structure, and are non-edible, positioning them as promising feedstock candidates. They can be cultivated effortlessly in saline or wastewater environments and aid in carbon dioxide sequestration. Despite the complex digestibility of macroalgal cell walls, their lipid and carbohydrate content renders them ideal for bioethanol and biodiesel synthesis (Mohammad et al. 2019).

Biofuel manufacturers face several challenges, such as the volatile market for unrefined vegetable oil derived from edible plants, the supply of grains, and the timely acquisition of raw materials. Collectively, these factors impact the manufacturing of first-generation biofuels, which account for approximately 70%–90% of production. Furthermore, the bioenergy sector competes directly with food producers, a dynamic widely regarded as a pivotal driver of escalating food costs (Lai et al., 2005) To render biofuel production economically viable and environmentally sustainable, while mitigating conflicts with the food industry, enterprises must prioritize the development of second-generation biofuels derived from alternative, economically feasible feedstock sources such as biomass, waste oils and fats, and byproducts. Prospects like cellulosic ethanol and biodiesel obtained from waste oils, non-edible crops, or algae present tangible remedies to this quandary (Rodionova et al., 2017).

10.5 PRODUCTION PROCESSES

Biofuel production or processing can involve various technologies, including: physical technologies, chemical technologies, thermochemical technologies, and biochemical technologies (Figure 10.3). Often, a combination of these technologies is employed (Teklit et al., 2021).

10.5.1 Physical extraction technologies

Certain physical technologies can be utilized for the extraction of specific biofuels. Mechanical approaches may be utilized on oil crops to acquire biooil, which is then separated from impurities through the process of distillation (Divjot et al., 2019).

10.5.1.1 Extraction process

Oil-rich rapeseed undergoes physical extraction through mechanical crushing, followed by the transformation of the oil into esters, which serve as direct substitutes for diesel. Commonly used mechanical pressure methods, such as screw presses, are known for their effectiveness in prepressing, leading to a higher oil yield from the seeds (Peng et al, 2019).

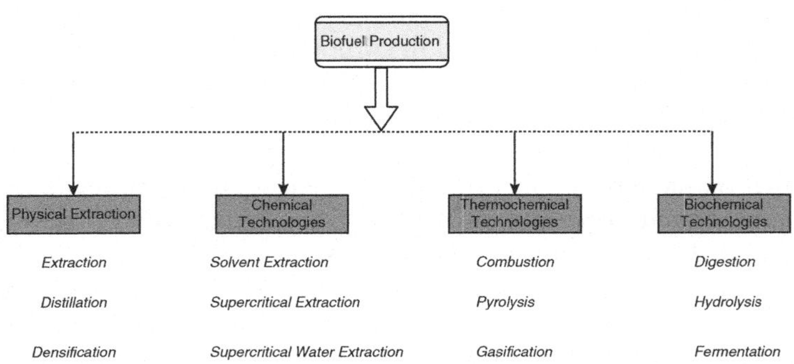

Figure 10.3 Various technologies in biofuel production.

10.5.1.2 Distillation method

Various methodologies of distillation exist, including molecular distillation and vacuum distillation, wherein the vaporization of essential oils occurs alongside steam, ultimately resulting in the collection of biooil. This phase is executed after the mechanical crushing process. The property of volatility is of paramount significance in this context, as it determines the effective separation of the diverse constituents within the liquid mixture (Christine et al., 2010).

10.5.1.3 Densification techniques

The byproducts generated from agricultural and forestry activities present significant challenges for conversion into fuel sources. Their inherent characteristics, including varied particle size, low bulk density, and minimal energy content, necessitate a series of pre-processing steps, such as cutting, crushing, and pulverization, to facilitate their transformation into high-quality fuel alternatives suitable for practical utilization (Sarker et al., 2003).

10.5.2 Chemical technologies

Solvent extraction, supercritical fluid extraction, and supercritical water conversion represent the predominant chemical methodologies used in the sphere of biofuel production (Bhatt et al., 2023).

10.5.2.1 Solvent extraction

The selection of feedstock, such as aromatic grasses, wheat straw, and wood, deemed suitable for this process, is based on the presence of specific valuable primary and secondary metabolites. An appropriate solvent (acetone, hexane, ethanol, and isopropanol) is employed to facilitate the extraction of oil from oilseeds, after which the utilized solvent is subjected to evaporation. The oil–solvent mixture undergoes distillation to achieve optimal purification. The resulting de-oiled meal is subsequently subjected to toasting (Efthymiopoulos et al., 2019).

10.5.2.2 Supercritical extraction

This procedure relies on two fundamental parameters—temperature and pressure—to facilitate the conversion of biomass such as cedar wood, sandalwood, and pine wood into valuable biofuels. Carbon dioxide is used to extract secondary metabolites, including terpenoids, waxes, resins, sterols, polycosanols, and alkaloids, from the biomass. This phase is followed by hydrolysis and fermentation processes, wherein the residual lignocellulosic biomass undergoes hydrolysis and fermentation, ultimately yielding the biofuels (Akalın et al., 2017).

10.5.2.3 Supercritical water conversion

This method involves subjecting biomass to water within a temperature range of 300–644 K and a pressure of 200–250 bar. The dissociated water into acidic and basic components permeates the biomass, leading to the breakdown of cellulose and hemicellulose bonds and the production of monosaccharides and oligosaccharides. Notably, this system eliminates the need for a catalyst (Vogel et al., 2007).

10.5.3 Thermochemical technologies

Several versatile thermochemical procedures have been reported that aim to transform the organic components of the feedstock into a variety of products, predominantly biofuels (Virmond et al., 2013).

10.5.3.1 Combustion

Combustion is a phenomenon in which a fuel undergoes oxidation in an environment abundant in oxygen, liberating the latent chemical energy in the form of thermal energy. This generated heat finds application in culinary activities, indoor climate control, and power generation. Although all forms of biomass or biofuels have the potential for complete combustion, ensuring optimal combustion necessitates maintaining the moisture content of the fuel below the threshold of 50%. Often, a preliminary heating process is employed to guarantee compliance with this criterion (Kohse-Höinghaus. et al., 2010)

10.5.3.2 Pyrolysis

Pyrolysis denotes a thermochemical transformation that disintegrates biomass raw material at temperatures ranging from moderate (300°C–800°C) to elevated (800°C–1,300°C) within a setting devoid of reactive gases. The resultant byproducts include biooil, biosyngas, and biochar. Diverse categories of pyrolysis encompass gradual pyrolysis, rapid pyrolysis, instantaneous pyrolysis, catalytic pyrolysis, microwave pyrolysis, vacuum pyrolysis, and hydropyrolysis. Among these, microwave-assisted pyrolysis (MAP) emerges as a recently developed technique garnering attention in contemporary scholarly works and applications (Demirbas, 2008).

Microwave-assisted pyrolysis and traditional electrical pyrolysis represent two distinct methodologies for thermal degradation procedures, each presenting unique merits and considerations. Microwave-assisted pyrolysis leverages non-contact heating via electromagnetic radiation, ensuring homogeneous and volumetric heating across the material. This approach excels in prompt heating and shortened reaction durations owing to reduced thermal resistance, thereby enhancing energy efficiency and product output. Conversely, conventional electrical pyrolysis relies on direct contact heating, resulting in gradual heating rates and prolonged reaction periods due to heightened thermal resistance. Although it enables superficial heating and is more universally adoptable in diverse configurations, it typically entails greater energy consumption and provides less control over the heating process, culminating in inferior product yields in comparison to microwave-assisted pyrolysis. Consequently, whereas microwave-assisted pyrolysis places a premium on efficiency and careful regulation, traditional electrical pyrolysis offers adaptability, albeit with trade-offs in energy utilization and heating effectiveness (Bridgwater, 2003).

10.5.3.3 Gasification

Gasification is a chemical process that involves the reaction of fuel with a gasification agent to generate syngas comprising CO, H_2, N_2, CO_2, and some hydrocarbons. Numerous gasification techniques are available, among which air gasification stands out as the most extensively studied and applied. The popularity of air gasification can be attributed to the cost-effectiveness of the gasification agent (air), the simplicity of the reaction mechanism, and the uncomplicated design of the reactor. Generally characterized as endothermic, the gasification process necessitates continuous external heating throughout its operation (Molino et al., 2018).

A gasifier is the apparatus in which the gasification reactions occur. These gasifiers play a vital role in transforming biomass or other raw materials into usable gas products, with each category offering distinct benefits based on specific application needs. Fixed bed gasifiers, suitable for capacities within the range of 0.01–10 MW, are adept at handling large particles and function at relatively low temperatures (450°C–650°C). Nonetheless, they tend to produce gas streams with high levels of particulate matter and require significant quantities of gasifying agents, potentially resulting in elevated tar content in the gas (up to 150 g/Nm^3). On the contrary, fluidized bed gasifiers, ranging in capacities from 1 to 100 MW, provide a more consistent temperature distribution and improved gas–solid interaction due to their operating temperatures of 1000°C–1,200°C. They are well-suited for feedstocks with lower ash fusion temperatures and yield gas streams with reduced particulate content. The elimination of ash in both fixed and fluidized bed gasifiers is typically handled as either slag or dry ash. Conversely, entrained flow gasifiers, engineered for large capacities between 60 and 1,000 MW, necessitate finely divided feed materials and operate at extremely high temperatures (>1,200°C). While they offer short residence times and effective slag elimination, they require substantial oxygen input and are less appropriate for feedstocks with high ash content. Therefore, each type of gasifier exhibits a distinctive array of features tailored to specific operational scales and feedstock compositions, influencing their applicability across diverse industrial sectors (Takuya et al., 2004).

10.5.4 Biochemical technologies

Biochemical processes for the conversion of biomass primarily encompass digestion, fermentation, and hydrolysis techniques. In some instances, relevant literature also references interesterification and transesterification (Abelniece. et al. 2022). In practical scenarios, these methods are frequently combined synergistically to enhance the efficient production of biofuels.

10.5.4.1 Digestion

The biochemical conversion of biomass through anaerobic or aerobic conditions, known as biomass digestion, employs bacteria to break down the biomass feedstock. Anaerobic digestion presents various advantages over aerobic digestion, such as elevated organic removal efficiency, a higher organic loading rate, reduced nutrient demands, and decreased energy consumption. Given that aerobic digestion might not efficiently convert biomass feedstock into valuable end products, anaerobic digestion prevails in usage (Guiot and Frigon, 2012).

Anaerobic digestion denotes the biochemical transformation of biomass by diverse bacteria in the absence of oxygen, resulting in the production of biogas comprising mainly 55%–75% methane and 25%–45% carbon dioxide, alongside trace gases like hydrogen sulfide. This process entails four primary stages: hydrolysis, acidogenesis, acetogenesis, and methanogenesis. It is a multifaceted process applicable to a range of biomass feedstocks with high moisture content, including municipal solid waste, industrial waste, livestock manure, food processing waste, and domestic and industrial sewage (Torres et al., 2023).

10.5.4.2 Hydrolysis

In the process of converting lignocellulosic biomass to bioethanol, the initial step involves hydrolyzing the biomass into fermentable sugars, followed by fermentation to produce alcohol. Hydrolysis represents a fundamental phase that influences the overall efficiency of the process. Chemical and enzymatic methodologies are the preferred strategies for sugar extraction from lignocellulosic biomass (Parameswaran et al., 2011).

Mineral acids are considered cost-effective for chemical hydrolysis. Nevertheless, the low yield obtained has prompted researchers to explore environmentally friendly cellulose solvents like ionic liquids. Enzymes have also been utilized for hydrolysis due to their specificity, minimal side products, and corrosion-free process. However, these enzymes are costly and vulnerable to inhibition by the final product (Vasić et al., 2021).

10.5.4.3 Fermentation

Grains, timber, sugar crops, and starch crops are examples of biomass that undergo fermentation for large-scale bioethanol production. The initial step involves a hydrolysis procedure for sugar extraction, followed by fermentation to yield bioethanol. Yeast and bacteria play crucial roles in the fermentation pathway, resulting in the formation of alcohol and other side products. High-quality ethanol with a high octane rating is acquired through two main processes: distillation and dehydration (Gustavo et al., 2008).

10.6 SUSTAINABILITY CONCERNS

Biofuels, originating from organic matter, are considered renewable energy sources that are known for their positive impact on the environment. Unlike fossil-based fuels, biofuels are associated with decreased emissions of air pollutants, thereby contributing to reduced greenhouse gas emissions. An illustrative example is the diminished production of nitrous oxide (N_2O) resulting from the utilization of biodiesel as opposed to conventional diesel fuel. This marks a significant advancement in addressing climate change (Ayhan, 2009).

The reduced emissions from biofuels can be attributed to their elevated oxygen content, which ranges from 10% to 45%, enhancing combustion efficiency. This efficiency can be further improved by blending biofuels with oxygenates such as ethanol and methanol, consequently reducing the release of unburned hydrocarbons into the atmosphere (Joseph et al., 2010).

To boost the production of biofuels, it is essential to implement workable agricultural practices. The quantity of biofuels needed for extensive applications in the future heavily relies on the utilization of vast agricultural and forest lands, which can have detrimental effects on soil fertility, necessitate the use of synthetic fertilizers, consume significant amounts of water, and impact biodiversity. This can further pose adverse consequences for soil organisms (Fernando and Suani, 2017).

Agricultural residues and non-edible biomass serve as superior alternatives to agricultural and forest resources needed for the production of biofuels. Such a substitution positively influences biodiversity, land utilization challenges, and the carbon content of soils. Addressing this issue may concurrently alleviate concerns associated with heightened water consumption in regions characterized by water scarcity (Jeswani et al., 2020).

An additional strategy to rehabilitate soil structure involves the practice of phytoremediation, alongside the cultivation of rapidly growing indigenous plant species (Ben, 2009).

The dynamics of biofuel consumption exhibit significant variation across different nations; for instance, developing countries predominantly emerge as the principal consumers of biofuels. This phenomenon is contingent upon factors such as land availability, environmental repercussions, production costs, and consumption behaviors (Yogeeswari and Tajul, 2021).

Cultural impediments, a deficiency of a proficient labor force, and insufficient infrastructure serve as detrimental factors regarding the production of biofuels, whereas favorable policies related to agriculture contribute positively to its advancement. Various flora, including selected species of plants and shrubs, have been identified in scholarly literature as particularly advantageous for biofuel production, such as palm oil, jatropha (Al-Khayri et al., 2022), and cassava (Christer et al., 2009). Should these species be cultivated extensively for biofuel generation, it may result in significant deforestation. This issue deserves careful consideration (Arora et al., 2020).

It is imperative to establish a balance among production expenditures, distribution costs, and initial capital outlay. However, ongoing operational expenses have surpassed the initial investments (Arora et al., 2021). The prospects for biofuels appear promising only when they follow a trajectory devoid of subsidies (Wu and Langpap, 2015).

10.7 APPLICATIONS OF BIOFUELS

The value of bioethanol and biodiesel in the realm of biofuels stems from their role in maintaining favorable climatic conditions and reducing greenhouse gas emissions. However, upon analysis, it is evident that the collection of biomass, along with its processing, transportation, and distribution, significantly contributes to greenhouse gas emissions. Hence, these factors must be taken into account before reaching a conclusive assessment regarding the overall importance of biofuels (Hui et al., 2022).

The expansion of the biofuels market significantly boosts agricultural activities, providing an opportunity to cultivate a wide range of crops to meet both national and international energy needs. Over time, reliance on subsidies is likely to decrease as countries with agricultural surpluses experience a severe decline in farm income. Underserved rural areas undergo revitalization, leading to incremental job creation (Jikun et al., 2012).

In rural areas of Southeast Asia and sub-Saharan Africa, women engage in labor-intensive tasks such as collecting biomass. Traditional cooking methods utilizing wood, dung, and crop residues have resulted in instances of respiratory illnesses and premature deaths due to inhalation of polluted air. The successful implementation of biogas can thus improve women's health and relieve them of the burdens associated with collecting materials for fuel production (Tandon, 2009).

The methyl ester present in biofuels is recognized for its potent solvent properties, facilitating the cleaning of contaminated coastlines. This method is environmentally friendly in addressing oil spills, thereby contributing to the promotion of environmental sustainability and positioning itself in opposition to conventional diesel and oil-based lubricants (Hájek et al., 2021).

The ecological influence of non-biodegradable plastic waste is immeasurable, as millions of tons of plastics are produced annually. Numerous microorganisms have shown promise in plastic degradation. This ongoing process transforms plastics into biofuels. For instance, *Ideonella sakaiensis* can decompose polyethylene terephthalate (PET) into terephthalic acid and ethylene glycol through enzymatic pathways. Additionally, various bacteria and fungi are known to degrade polyurethane and polystyrene. Although enzymatic pathways for some plastics remain limited, non-enzymatic treatments like pyrolysis effectively convert plastics into long-chain hydrocarbons. These hydrocarbons can be utilized by microorganisms such as *Yarrowia lipolytica*, which can further metabolize them into valuable fuels and chemicals. By harnessing and integrating these microbial pathways into engineered hosts, we can facilitate plastic waste management and create sustainable biofuels and biodegradable alternatives, addressing both energy needs and environmental concerns (Mohammed, 2021).

10.8 BIOGAS PRODUCTION AND APPLICATIONS

Biogas constitutes a sustainable and renewable energy source with the potential to substitute traditional fossil fuels, especially in rural regions. The generation of biogas not only tackles the pressing ecological and environmental issues linked to the depletion of fossil fuels but also utilizes organic waste materials. It mainly consists of methane (CH_4, 55%–70%) and carbon dioxide (CO_2, 30%–45%), alongside small quantities of hydrogen sulfide (H_2S), among other elements. Methane, being the primary constituent, holds significant value for processes such as fueling vehicle engines, despite being a competent renewable energy source. Nonetheless, the release of methane into the atmosphere poses notable risks associated with greenhouse gas emissions (Kougias and Angelidaki, 2018).

The predominant technique for producing biogas from organic substances, such as manure, sewage sludge, municipal solid waste, biodegradable waste, and agricultural slurry, is anaerobic digestion. This technique is an efficient strategy that converts common non-marketable substances into valuable energy sources (Munish et al., 2012).

Biogas initiation occurs in anaerobic reactors of several capacities ranging from a minimum of 1–2000 m^3, involving a diversity of microorganisms. Acidogens play a critical role in breaking down the organic material into simpler compounds such as monomers and dimers. The methane gas produced is directly proportional to the quality of the organic material used during the process. The initial stage is considered the rate-limiting step since it encompasses the dissolution of solid particles. A group of microorganisms known as methanogens utilizes the monomers and dimers from the first step, ultimately resulting in the emission of biogas. The acceleration of anaerobic digestion can be achieved by incorporating certain mechanical or thermal treatments just prior to the commencement of the process. Following biogas generation, there is a discharge of undigested fiber along with water-soluble components referred to as digestate (a good fertilizer), which serves as a byproduct (Chunlan et al., 2015).

The process of anaerobic digestion is highly dependent on the carbon-to-nitrogen ratio within agricultural residues. A decrease in this ratio results in reduced rates of gas production. This ratio can be regulated with the help of nitrogen sources, leading to increased biogas production (Tanimu et al., 2014).

10.8.1 Types of digesters

Biogas production depends on several factors like soil composition, temperature, groundwater levels, and the feedstock utilized. Batch flow, continuous flow, continuously expanding, plug flow, and contact flow represent fundamental types of digesters (Rajendran et al., 2012).

Batch digesters involve a series of interconnected tanks. The organic matter is introduced simultaneously into the digester, where it undergoes fermentation over a period of time before being extracted. After each cycle, a fresh lot of feedstock is added to replace the digested material. Conversely, continuous digesters necessitate a daily intake of feedstock, with each batch requiring a retention period ranging from 15 to 40 days, subject to adjustments through agitation and heating variations. Despite the modest amount of diluted raw material utilized in individual units, a consistent output of biogas and biofertilizer is generated. Occasionally, a tandem operation involving two digesters is implemented, focusing on gas production in the initial stage while the latter digester facilitates the gradual completion of the digestion process. The progressive filling technique is adopted in continuously expanding digesters, commencing at one-third capacity and incrementally reaching full capacity before being emptied (Parawira et al., 2004). Notably, in plug-flow digesters, feedstock introduction occurs at one end, traversing through to the opposite end, with the indispensable presence of a support medium in contact flow digesters (Lili et al., 2019).

The configuration of biogas digesters varies depending on the location, substrate availability, and climate conditions. In mountainous regions, digesters are designed with reduced gas volume to minimize gas loss, while tropical countries prefer underground digesters to harness geothermal energy. Among the diverse construction models, the fixed dome model pioneered by China (Castano et al., 2014) and the floating drum model developed by India (Sathish and Vivekanandan, 2016) have demonstrated effectiveness and continue to be utilized to this day.

Chinese biogas production is conducted through fixed dome digesters, commonly known as "Chinese" or "hydraulic" digesters due to their design (Figure 10.4). The longevity of these digesters is contingent upon their placement below the surface. The introduction of feedstock into the digesters is carried out in slurry form until it reaches the same level as the base of the expansion chamber. The greater the biogas collection, the higher the pressure applied to the substrate, prompting it to transition into the expansion chamber. Upon the liberation of biogas, the slurry flows back to the main digester (Chenxi et al., 2021). The "Janta" and "Deenbandhu" models represent adapted versions of this model in the context of India. The Janta model, established in 1978, is characterized by a shallow well design with a dome roof. However, it is limited by restricted slurry circulation and lower gas production. In response to these challenges, Action for Food Production (AFPRO) introduced the Deenbandhu model in 1984, which consists of two spheres of different diameters to address the limitations of the Janta model's efficiency (Kalia and Kanwar, 1996). In alignment with these developments, numerous countries have customized their digesters based on the Chinese model.

The floating drum digester, innovated by the Khadi and Village Industries Commission (KVIC) in 1962, is widely embraced and practiced for household applications in India. This distinctive model incorporates a mobile

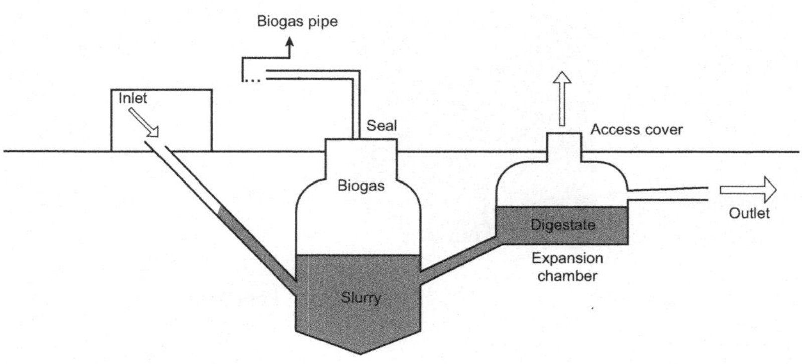

Figure 10.4 Schematic diagram of anaerobic digestion-fixed dome model.

inverted drum positioned atop a well-configured digester, with the drum serving as a storage reservoir that ascends and descends in response to gas accumulation levels. Leveraging the drum's weight, the requisite pressure for gas transmission through the pipeline is generated. Operationally, floating drum digesters deliver biogas at a consistent pressure level with variable volume, and the gas quantity can be readily determined based on the drum's position. Nonetheless, regular repainting of the drum is essential to avert rusting, and caution must be exercised to prevent fibrous materials from causing blockages (Sathish and Vivekanandan, 2016).

10.8.2 Pros of biogas

Biogas, the current sustainable energy source, has been integrated into urban transport operations in Sweden since 2004. It exhibits versatility in its application across engines such as spark ignition gas engines, gas turbines, fuel cells, and industrial heaters, among others. The economic viability of biogas is enhanced through its combined effect with heat and power systems (Rajmohan et al., 2021) (Figure 10.5).

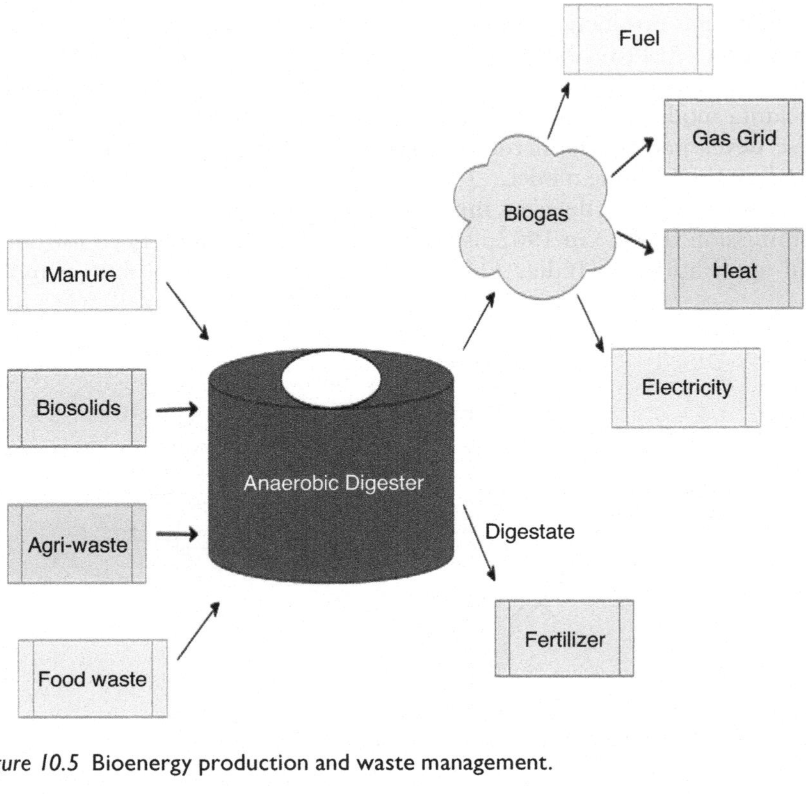

Figure 10.5 Bioenergy production and waste management.

In areas with high population densities and large landfill sites, the utilization of biogas technology emerges as a crucial factor in the management of municipal solid waste (Chukwunonso and Saeid, 2018). Furthermore, integrating biogas into various existing biorefinery models, such as bioethanol and biodiesel production, helps address the byproducts generated during these processes, consequently promoting sustainability within the energy sector (Beschkov, 2017).

10.8.3 Limitations

Biogas production occurs within large-scale digesters, necessitating a considerable investment. The duration that organic material remains in the digester exceeds 30 days, and this extended period can decrease efficiency, obstructing the production of biogas. In colder regions during winter, biogas output tends to be lower due to the direct connection between biogas production and temperature. Inadequate comprehension of the technology, irregular raw material supply, and the risk of gas seepage can lead users to abandon the digester and discontinue operations. Negligence increases the potential for explosions due to methane leaks in households, necessitating supervision (Muhammad et al., 2022).

A lack of technical expertise, absence of necessary raw materials, and inadequate monitoring pose barriers to the success of biogas technology in rural areas. Despite substantial technological advancements, non-governmental organizations are not fully engaged in disseminating these developments. A shortage of skilled workforce and initial funding could hinder a country's advancement, similar to the challenges faced by Thailand in the effective adoption of biogas technology. Providing fundamental knowledge and training to local communities is essential for success (Shivika et al., 2018).

Future prospects include advanced research in integrating biogas into vehicles. The proliferation of biogas refueling stations and distribution networks must increase proportionately to support the operation of biogas-powered vehicles. Dual-fuel vehicles, which run on both biogas and traditional fuels, incur high costs, presenting a barrier to experimenting with this technological breakthrough (Yousuf et al., 2016).

10.9 CONCLUSION AND FUTURE PROSPECTS

Biofuels emerge as a promising solution to mitigate environmental impacts from energy generation. However, improved agricultural practices and cultivation strategies are crucial for addressing potential barriers to the full deployment of biofuels. By weighing the advantages and disadvantages, biofuels can have a substantial impact on moving towards a more sustainable energy outlook. Second and potentially third-generation biofuels are more appealing from an economic perspective.

Bioethanol and biodiesel are poised to continue being prominent biofuels due to their well-established markets and advancements in technology. Nevertheless, their utilization is impeded by a lack of chemical diversity in comparison to conventional fuels. There is a necessity for upcoming biofuels to present heightened energy content and improved performance under low temperatures to enhance their feasibility. Moreover, there is a demand for biofuels customized for specific applications, such as in the aerospace industry. To accomplish these objectives, novel production techniques that capitalize on progress in synthetic biology are imperative, potentially enabling the development of biofuels with tailored attributes and enhanced manufacturing capabilities.

Biogas emerges as a highly promising renewable energy source, presenting a wide array of applications ranging from waste management to energy generation. The substantial capacity of biogas to reduce greenhouse gas emissions, coupled with its adaptability in energy generation, establishes it as a crucial element in the shift towards a more sustainable energy landscape. Biogas is firmly positioned as a key player in the transition towards a more sustainable and environmentally friendly energy framework.

REFERENCES

Abelniece Z., Laipniece L. and Kampars V. (2022) Biodiesel production by interesterification of rapeseed oil with methyl formate in presence of potassium alkoxides. *Biomass Conversion and Biorefinery,* **12**: 2881–2889. https://doi.org/10.1007/s13399-020-00874-z

Acheampong M., Ertem F.C., Kappler B. and Neubauer P. (2017) In pursuit of sustainable development goal (SDG) number 7: will biofuels be reliable? *Renewable and Sustainable Energy Reviews,* **75**: 927–937. https://doi.org/10.1016/j.rser.2016.11.074

Agarwal S. and Kumar A. (2018) Historical development of biofuels. In: *Biofuels: Greenhouse Gas Mitigation and Global Warming*, 17–45. https://doi.org/10.1007/978-81-322-3763-1_2

Akalın M.K., Tekin, K. and Karagöz, S. (2017) Supercritical fluid extraction of biofuels from biomass. *Environmental Chemistry Letters,* **15**: 29–41. https://doi.org/10.1007/s10311-016-0593-z

Al-Khayri J.M., Sudheer W.N., Preetha T.R., Nagella P., Rezk A.A. and Shehata W.F. (2022) Biotechnological research progress in jatropha, a biodiesel-yielding plant. *Plants (Basel),* **11**: 1292. https://doi.org/10.3390/plants11101292

Arora K., Kumar A., Kamboj V.K., Prashar D., Jha S., Shrestha B. and Joshi G.P. (2020) Optimization methodologies and testing on standard benchmark functions of load frequency control for interconnected multi area power system in smart grids. *MDPI Journal of Mathematics*, **8**(6): 980. https://doi.org/10.3390/math8060980

Arora K., Kumar A., Kamboj V.K., Prashar D., Shrestha B. and Joshi G.P. (2021) Impact of renewable energy sources into multi area multi-source load frequency control of interrelated power system. *MDPI Journal of Mathematics*, **9**(2): 186. https://doi.org/10.3390/math9020186

Awogbemi O., Kallon D.V.V., Onuh E.I. and Aigbodion V.S. (2021) An overview of the classification, production and utilization of biofuels for internal combustion engine applications. *Energies*, **14**: 5687. https://doi.org/10.3390/en14185687

Ayadi M., Saurabh J.S., Pachapur V., Brar S.K. and Ridha B.C. (2016) History and global policy of biofuels. In: *Green Fuels Technology*, 1–14. https://doi.org/10.1007/978-3-319-30205-8_1

Ayhan D. (2009) Political, economic and environmental impacts of biofuels: a review. *Applied Energy*, **86**: S108–S117. https://doi.org/10.1016/j.apenergy.2009.04.036

Ben P. (2009) The social and environmental impacts of biofuels in Asia: an overview. *Applied Energy*, **86**: S21–S29. https://doi.org/10.1016/j.apenergy.2009.04.046

Beschkov V. (2017) Biogas, biodiesel and bioethanol as multifunctional renewable fuels and raw materials. In: Jacob-Lopes E. and Queiroz Z.L. (Eds), *Frontiers in Bioenergy and Biofuels*, IntechOpen, London, 185–205. https://doi.org/10.5772/65734

Bhatt B, Sharma H., Arora K., Joshi G.P. and Shrestha B. (2023) Levy flight-based improved grey wolf optimization: a solution for various engineering problems. *MDPI Journal of Mathematics* **11**(7): 1745. https://doi.org/10.3390/math11071745

Bridgwater A.V. (2003) Renewable fuels and chemicals by thermal processing of biomass. *Chemical Engineering Journal*, **91**: 87–102, https://doi.org/10.1016/S1385-8947(02)00142-0

Carlo N.H. and André P.C.F. (2006) Outlook for advanced biofuels. *Energy Policy*, **34**: 3268–3283. https://doi.org/10.1016/j.enpol.2005.06.012

Castano J.M., Martin J.F. and Ciotola R. (2014) Performance of a small-scale, variable temperature fixed dome digester in a temperate climate. *Energies*, **7**: 5701–5716. https://doi.org/10.3390/en7095701

Chenxi J., Shiqiang S., Dianhai Y., Weijie S., Yadong M., Wenzhi H. and Guangming L. (2021) Anaerobic digestion: an alternative resource treatment option for food waste in China. *Science of the Total Environment*, **779**: 146397. https://doi.org/10.1016/j.scitotenv.2021.146397

Christer J., Anna W., Jiaming Z., Xinwen H. and Chuanxin S. (2009) Cassava, a potential biofuel crop in (the) People's Republic of China. *Applied Energy*, **86**: S95–S99. https://doi.org/10.1016/j.apenergy.2009.05.011

Christine B., Sigurd S. and Martin M. (2010) Simulated distillation for biofuel analysis. *Energy Fuels*, **24**: 2086–2090. https://doi.org/10.1021/ef901295s

Chukwunonso C.A. and Saeid B. (2018) Decentralized anaerobic digestion systems for increased utilization of biogas from municipal solid waste. *Renewable and Sustainable Energy Reviews*, **90**: 982–991. https://doi.org/10.1016/j.rser.2018.03.009

Chunlan M., Yongzhong F., Xiaojiao W. and Guangxin R. (2015) Review on research achievements of biogas from anaerobic digestion. *Renewable and Sustainable Energy Reviews*, **45**: 540–555. https://doi.org/10.1016/j.rser.2015.02.032

Demirbas A. (2008) Biofuels sources, biofuel policy, biofuel economy and global biofuel projections. *Energy Conversion and Management*, **49**: 2106–2116. https://doi.org/10.1016/j.enconman.2008.02.020

Divjot K., Kusam L.R., Neelam Y., Ajar N.Y., Ali A.R., Chhatarpal S., Puneet N., Karan S. and Anil K.S. (2019) Technologies for biofuel production: current development, challenges, and future prospects. In: Rastegari A., Yadav A. and Gupta A. (Eds), *Prospects of Renewable Bioprocessing in Future Energy Systems. Biofuel and Biorefinery Technologies*, vol. 10. Springer,Cham. https://doi.org/10.1007/978-3-030-14463-0_1

Efthymiopoulos, I., Hellier, P., Ladommatos, N., Kay A. and Ben M. (2019) Effect of solvent extraction parameters on the recovery of oil from spent coffee grounds for biofuel production. *Waste Biomass Valor,* **10**: 253–264. https://doi.org/10.1007/s12649-017-0061-4

Fernando C.D. and Suani T.C. (2017) History, evolution, and environmental impact of biodiesel in Brazil: a review. *Renewable and Sustainable Energy Reviews*, **75**: 168–179. https://doi.org/10.1016/j.rser.2016.10.060

Guiot, S.R. and Frigon, J.C. (2012). Anaerobic digestion as an effective biofuel production technology. In: Hallenbeck, P. (ed.), *Microbial Technologies in Advanced Biofuels Production*. Springer, Boston, MA. https://doi.org/10.1007/978-1-4614-1208-3_9

Günnur K. and Nilgün C. (2013) An overview of biofuels from energy crops: current status and future prospects, *Renewable and Sustainable Energy Reviews*, **28**: 900–916. https://doi.org/10.1016/j.rser.2013.08.022

Gustavo D., Felipe A., Antonio D.L. and Elias R. (2008) Fermentative hydrogen production in batch experiments using lactose, cheese whey and glucose: influence of initial substrate concentration and pH. *International Journal of Hydrogen Energy*, **33**(19): 4989–4997. https://doi.org/10.1016/j.ijhydene.2008.06.065

Gustavo de L.T.O., Ben M. and Christina P. (2017) How biofuel policies backfire: misguided goals, inefficient mechanisms, and political-ecological blind spots. *Energy Policy*, **108**: 765–775. https://doi.org/10.1016/j.enpol.2017.03.036

Hájek M., Vávra A., de P.C.H and Kocík J. (2021) The catalysed transformation of vegetable oils or animal fats to biofuels and bio-lubricants: a review. *Catalysts*, **11**: 1118. https://doi.org/10.3390/catal11091118

Hui X., Longwen O., Yuan L., Troy R.H. and Michael W. (2022) Life cycle greenhouse gas emissions of biodiesel and renewable diesel production in the United States. *Environmental Science & Technology*, **56**: 7512–7521. https://doi.org/10.1021/acs.est.2c00289

Indu S.T., Manish K., Sunita J.V., Yonghong W., Edgard G. and Sindhu R. (2018) Sequestration and utilization of carbon dioxide by chemical and biological methods for biofuels and biomaterials by chemoautotrophs: opportunities and challenges. *Bioresource Technology*, **256**: 478–490. https://doi.org/10.1016/j.biortech.2018.02.039

Jeswani H.K., Chilvers A. and Azapagic A. (2020) Environmental sustainability of biofuels: a review. *Proceedings of the Royal Society A*, **476**: 20200351. https://doi.org/10.1098/rspa.2020.0351

Jikun H., Jun Y., Siwa M., Scott R. and Alfons W. (2012) Biofuels and the poor: global impact pathways of biofuels on agricultural markets. *Food Policy*, **37**: 439–451. https://doi.org/10.1016/j.foodpol.2012.04.004

Joseph E.H., Richard J.P. and Jason D.H. (2010) The ecological impact of biofuels. *Annual Review of Ecology, Evolution, and Systematics*, **41**: 351–377. https://doi.org/10.1146/annurev-ecolsys-102209-144720

Kalia A.K. and Kanwar S.S. (1996) Performance evaluation of janata and deenbandhu biogas plants. *Energy Sources*, **18**(3): 283–289. https://doi.org/10.1080/00908319608908767

Kohse-Höinghaus K., Osswald P., Cool T.A., Kasper T., Hansen N., Qi F., Westbrook C.K. and Westmoreland P.R. (2010) Biofuel combustion chemistry: from ethanol to biodiesel. *Angewandte Chemie International Edition in English*, **49**: 3572–97. https://doi.org/10.1002/anie.200905335

Kougias P.G. and Angelidaki I. (2018) Biogas and its opportunities – a review. *Frontiers of Environmental Science and Engineering*, **12**: 14. https://doi.org/10.1007/s11783-018-1037-8

Lai C.C., Zullaikah S., Vali S.R. and Ju Y.H. (2005) Lipase-catalyzed production of biodiesel from rice bran oil. *Journal of Chemical Technology and Biotechnology*, **80**: 331–337. https://doi.org/10.1002/jctb.1208

Lili D., Guangli C., Xianzhang G., Tianshu L., Jiwen W. and Nanqi R. (2019) Efficient biogas production from cattle manure in a plug flow reactor: a large scale long term study. *Bioresource Technology*, **278**: 450–455. https://doi.org/10.1016/j.biortech.2019.01.100

Michael B.C., Rachel R., Neal R. and Richard O. (2007) Public policy and biofuels: the way forward? *Energy Policy*, **35**: 5737–5746. https://doi.org/10.1016/j.enpol.2007.06.008

Mingxin G., Weiping S. and Jeremy B. (2015) Bioenergy and biofuels: history, status, and perspective. *Renewable and Sustainable Energy Reviews*, **42**: 712–725. https://doi.org/10.1016/j.rser.2014.10.013

Mohammad H.K., Ismail E., Jose M.L., Fabienne R., R., Francisco J.B., James C. and Amin M.K. (2019) Advances in plant materials, food by-products, and algae conversion into biofuels: use of environmentally friendly technologies. *Green Chemistry*, **21**: 3213–3231. https://doi.org/10.1039/C8GC03860K

Mohammed B.A.R. (2021) Recent advancements of thermochemical conversion of plastic waste to biofuel: a review. *Cleaner Engineering and Technology*, **2**: 100062. https://doi.org/10.1016/j.clet.2021.100062

Molino A., Larocca V., Chianese S. and Musmarra D. (2018) Biofuels production by biomass gasification: a review. *Energies*, **11**: 811. https://doi.org/10.3390/en11040811

Morshadul H., Mohammad Z.A., Mohamamd B.A., Nekmahmud M. and Judit O. (2023) Sustainable biofuel economy: a mapping through bibliometric research. *Journal of Environmental Management*, **336**: 117644. https://doi.org/10.1016/j.jenvman.2023.117644

Muhammad U.K., Muhammad U., Muhammad A.A., Nalok D., Gang L. and Shicheng Z. (2022) A review of recent advancements in pretreatment techniques of lignocellulosic materials for biogas production: opportunities and limitations. *Chemical Engineering Journal Advances*, **10**: 100263. https://doi.org/10.1016/j.ceja.2022.100263

Munish P., Reinu E.A. and Colin J.B. (2012) Biofuel production: prospects, challenges and feedstock in Australia. *Renewable and Sustainable Energy Reviews*, **16**: 6022–6031. https://doi.org/10.1016/j.rser.2012.06.025

Tanimu, M.I., Mohd Ghazi T.I., Harun R.M. and Idris A. (2014) Effect of carbon to nitrogen ratio of food waste on biogas methane production in a batch mesophilic anaerobic digester. *International Journal of Innovation, Management and Technology*, **5**: 116–119. https://doi.org/10.7763/IJIMT.2014.V5.497

Parameswaran B., Janu K.U., Raveendran S. and Ashok P. (2011) Chapter 10- hydrolysis of lignocellulosic biomass for bioethanol production. In: Ashok P., Christian L., Steven C.R., Claude-Gilles D.and Edgard G. (Eds), *Biofuels*, Academic Press, 229–250. https://doi.org/10.1016/B978-0-12-385099-7.00010-3

Parawira W., Murto M., Zvauya R. and Mattiasson B. (2004) Anaerobic batch digestion of solid potato waste alone and in combination with sugar beet leaves. *Renewable Energy*, **29**: 1811–1823. https://doi.org/10.1016/j.renene.2004.02.005

Peng L., Kiyoshi S. and Hisao M (2019) Extraction techniques in sustainable biofuel production: a concise review. *Fuel Processing Technology*, **193**: 295–303. https://doi.org/10.1016/j.fuproc.2019.05.009

Rafael L., Lorenzo H., Juan M.C., James H.C., Jose M.H., Diego L., Jose M.M. and Antonio A.R. (2008) Biofuels: a technological perspective. *Energy & Environmental Science*, **1**: 542–564. https://doi.org/10.1039/B807094F

Rajendran K., Aslanzadeh S. and Taherzadeh M.J. (2012) Household biogas digesters – a review. *Energies*, **5**: 2911–2942. https://doi.org/10.3390/en508291

Rajmohan K., Ramya C. and Varjani S. (2021) Trends and advances in bioenergy production and sustainable solid waste management. *Energy & Environment*, **32**(6): 1059–1085. https://doi.org/10.1177/0958305X19882415

Ravi P., Sonil N. and Ajay K.D. (2021) Conversion of municipal solid waste to biofuels. In: Sonil N. and Dai-Viet N.V. (Eds), *Progressive Thermochemical Biorefining Technologies*, CRC Press, Boca Raton. https://doi.org/10.1201/9781003098591

Rodionova M.V., Poudyal R.S., Tiwari I., Voloshin R., Zharmukhamedov SK, Nam H.G., Bolatkhan Z., Bruce B.D., Hou, H.J.M. and Allakhverdiev S.I. (2017). Biofuel production: Challenges and opportunities. *International Journal of Hydrogen Energy,* **42**: 1–12. https://doi.org/10.1016/j.ijhydene.2016.11.125

Roger R., Yaning Z., Paul C., Shiyu L., Liangliang F., Nan Z., Kuan D., Peng P., Min A., Yanling C., Erik A., Yunpu W., Yuhuan L., Hanwu L. and Bingxi L. (2019) Chapter 1- biofuels: introduction. In: Ashok P., Christian L., Claude-Gilles D., Edgard G., Samir K.K. and Steven R. (Eds), *Biomass, Biofuels, Biochemicals, Biofuels: Alternative Feedstocks and Conversion Processes for the Production of Liquid and Gaseous Biofuels*, 2nd edition, Academic Press, 3–43. https://doi.org/10.1016/B978-0-12-816856-1.00001-4

Sakthivel R., Abbhijith H., Harshini G.V., Musunuri S.V. and Krushna P.S. (2023) Chapter 1- biofuels: classification, conversion technologies, optimization techniques and applications. In: Prakash K.S. (Ed.), *Biorefinery Production of Fuels and Platform Chemicals,* Wiley & Sons Ltd, 1–30. https://doi.org/10.1002/9781119724872.ch1

Sarker T.R., Nanda S., Meda V. and Dalai A.K. (2023) Densification of waste biomass for manufacturing solid biofuel pellets: a review. *Environmental Chemistry Letters,* **21**: 231–264. https://doi.org/10.1007/s10311-022-01510-0

Sathish S. and Vivekanandan S. (2016) Parametric optimization for floating drum anaerobic bio-digester using response surface methodology and artificial neural network. *Alexandria Engineering Journal*, **55**: 3297–3307. https://doi.org/10.1016/j.aej.2016.08.010

Shivika M., Erik O.A. and Shukla P.R. (2018) Barriers to biogas dissemination in India: a review. *Energy Policy*, **112**: 361–370, https://doi.org/10.1016/j.enpol.2017.10.027

Singh P., Arora K., Rathore U.C., Joshi G.P. and Cho W. (2024) Comparative study of controllers in battery energy storage system integrated with doubly fed induction generator-based wind energy conversion system for power quality improvement. *Energy Reports*, **11**(9): 4587–4600. https://doi.org/10.1016/j.egyr.2024.04.020

Takuya Y., Yoshito O. and Yukihiko M. (2004) Gasification of biomass model compounds and real biomass in supercritical water. *Biomass and Bioenergy*, **26**: 71–78. https://doi.org/10.1016/S0961-9534(03)00063-1

Tandon N. (2009) The bio-fuel frenzy: what options for rural women? A case of rural development schizophrenia. *Gender & Development*, **17**: 109–124. https://doi.org/10.1080/13552070802696961

Teklit G.A., Mentore V., Adrián B., Shiv P., Eric D.H. and Sami R. (2021) Emerging technologies for biofuel production: a critical review on recent progress, challenges and perspectives. *Journal of Environmental Management*, **290**: 112627. https://doi.org/10.1016/j.jenvman.2021.112627

Torres A., Padrino S., Brito A. and Diaz L. (2023) Biogas production from anaerobic digestion of solid microalgae residues generated on different processes of microalgae-to-biofuel production. *Biomass Conversion and Biorefinery*, **13**: 4659–4672. https://doi.org/10.1007/s13399-021-01898-9

Vasić K., Knez Ž. and Leitgeb M. (2021) Bioethanol production by enzymatic hydrolysis from different lignocellulosic sources. *Molecules*. **26**: 753. https://doi.org/10.3390/molecules26030753

Virmond E., Rocha J.D., Moreira R.F.P.M. and José H.J. (2013) Valorization of agroindustrial solid residues and residues from biofuel production chains by thermochemical conversion: a review, citing Brazil as a case study. *Brazilian Journal of Chemical Engineering*, **30**: 197–229. https://doi.org/10.1590/S0104-66322013000200001

Vogel F., Waldner M.H., Rouff A.A. and Rabe, S. (2007). Synthetic natural gas from biomass by catalytic conversion in supercritical water. *Green Chemistry*, **9**: 616–619. https://doi.org/10.1039/b614601e

Williams C.L., Westover T.L., Emerson, R.M., Tumuluru J.S. and Li C. (2016) Sources of biomass feedstock variability and the potential impact on biofuels production. *BioEnergy Research*, **9**: 1–14. https://doi.org/10.1007/s12155-015-9694-y

Wu J. and Langpap C. (2015) The price and welfare effects of biofuel mandates and subsidies. *Environmental and Resource Economics*, **62**: 35–57. https://doi.org/10.1007/s10640-014-9814-8

Yogeeswari S. and Tajul A.M. (2021) The impact of economic globalization on biofuel in developing countries. *Energy Conversion and Management: X*, **10**: 100064. https://doi.org/10.1016/j.ecmx.2020.100064

Yousuf A., Khan M.R., Pirozzi D. and Ab W.Z. (2016). Financial sustainability of biogas technology: barriers, opportunities, and solutions. *Energy Sources. Part B: Economics, Planning, and Policy*, **11**(9), 841–848. https://doi.org/10.1080/15567249.2016.1148084

Zul I. (2022) Chapter 3- Biomass classification and characterization for conversion to biofuels. In: Suzana Y. and Nor A.R. (Eds), *Value-Chain of Biofuels*, Elsevier, 69–87. https://doi.org/10.1016/B978-0-12-824388-6.00014-2

Chapter 11

Internet of Things (IoT)-enhanced fault detection and renewable integration for sustainable urban power systems

Vinay Anand and Himanshu Sharma

11.1 INTRODUCTION

The ever-increasing demands of modern urban infrastructure have necessitated a shift from overhead transmission lines to underground cables. This transition is primarily driven by the numerous advantages that underground cables offer, including reduced line losses, lower maintenance costs, and enhanced resilience against severe weather conditions, which are critical in ensuring the reliability and efficiency of power distribution systems in densely populated urban areas [1]. However, the deployment of underground cables introduces its own set of challenges, the most significant being the difficulty of accurately locating faults when they occur. Unlike overhead lines, where faults are often visible and can be quickly identified, underground cables require sophisticated detection techniques to pinpoint the exact location of faults, thereby minimizing downtime and maintenance costs [2].

In recent years, advancements in technology have paved the way for more innovative solutions to address these challenges [3]. Among these advancements, the integration of the Internet of Things (IoT) has shown promising potential. IoT-based systems offer real-time monitoring and fault detection capabilities, which can significantly enhance the accuracy of fault localization in underground cables. This study proposes an IoT-based fault detection system utilizing the Node Micro Controller Unit with Wireless Fidelity (WiFi) capabilities, combined with a Google database, to improve the efficiency and accuracy of fault detection in underground power cables [4].

In addition to fault detection in underground cables, the study also explores the integration of renewable energy sources into urban infrastructure, particularly focusing on photovoltaic (PV) systems and roadside wind turbines. The adoption of PV systems to drive induction motors in electric vehicles (EVs), as shown in Figure 11.1, is a critical step towards reducing the carbon footprint of urban transportation [5].

DOI: 10.1201/9781003542568-11

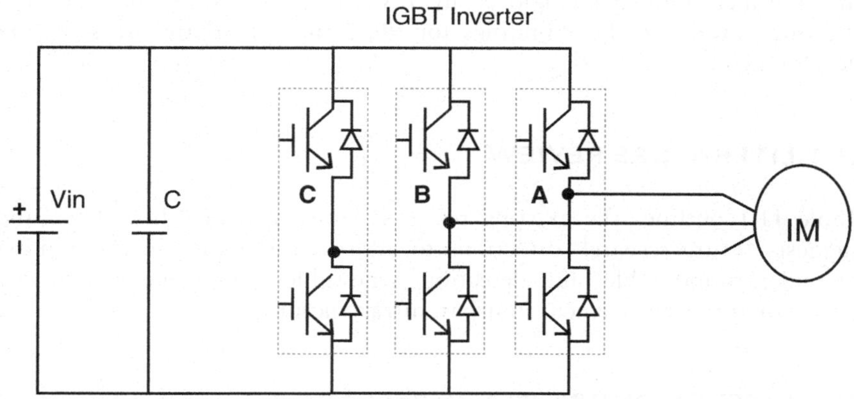

Figure 11.1 Induction motor-based electric vehicle drive.

This approach not only enhances the efficiency of EVs by minimizing voltage ripple but also contributes to the broader goal of sustainable energy utilization. The MATLAB/Simulink model used in this study demonstrates that incorporating PV systems into EVs can lead to reduced component sizes and costs, as well as decreased stress on switching devices, which are crucial for the long-term reliability and performance of EVs [6].

Furthermore, the study investigates the potential of roadside wind turbines to harness wind energy generated by moving vehicles. This energy can be stored in batteries for later use, thereby contributing to sustainable power generation. The integration of such renewable energy sources into urban infrastructure supports the global push towards greener energy solutions and provides a means to address the growing energy demands of modern cities [7].

To ensure the reliability of power systems, particularly those involving induction motors, the study also delves into the use of active power filters for power factor correction. Induction motors, widely used in various industrial and transportation applications, often face power quality issues such as voltage fluctuations and harmonic distortions. These issues can adversely affect the performance and efficiency of the motors. By implementing active power filters, the study aims to mitigate these problems, thereby improving the overall power quality and efficiency of the systems [8].

The research presented in this paper is structured around three key areas: (1) enhancing fault detection in underground cables through IoT-based systems, (2) integrating renewable energy sources such as PV systems and roadside wind turbines into urban infrastructure, and (3) improving power quality in induction motors using active power filters. Each of these areas contributes to the overarching goal of creating more reliable, efficient, and sustainable urban power systems. The following sections will delve deeper into the methodologies employed in this study, the results obtained, and the implications of these findings for the future of urban infrastructure development.

11.2 LITERATURE REVIEW

Table 11.1 outlines the existing research landscape, identifies gaps, and suggests future research directions to address these gaps in the context of underground cable fault detection, renewable energy integration, and power quality improvement in urban infrastructure.

11.3 EXISTING WORK IN THE RESEARCH

This research introduces a comprehensive approach that addresses several key areas within urban power systems, including underground cable fault detection, renewable energy integration, and power quality improvement. Below, Table 11.2 compares this research with existing works in these areas:

11.3.1 Enhanced fault detection

By integrating IoT technology, this research significantly improves the accuracy and real-time capabilities of fault detection in underground cables, addressing a critical gap in existing methods.

11.3.2 Renewable energy integration

The research advances the application of PV systems in EVs, as shown in Figure 11.2, and explores the novel use of roadside wind turbines shown in Figure 11.3, contributing to sustainable urban power systems [17].

11.3.3 Improvement of power quality

The study extends the application of active power filters to urban settings, providing solutions to power quality issues that are essential for modern urban infrastructure.

11.4 CHALLENGES AND CONSIDERATIONS

In advancing urban power systems, particularly with the integration of IoT-based fault detection, renewable energy sources, and power quality improvement methods, several challenges and considerations must be

Table 11.1 Outlines of the recent research

Research focus	*Authors and years*	*Key findings*	*Current gaps*	*Future directions*
Fault detection in underground cables	[9,10]	• Proposed a time-domain reflectometry (TDR) method for fault detection in underground cables. • Developed a machine learning model for fault prediction in power cables, improving fault detection speed and reliability	• TDR methods are limited by accuracy in complex network systems and are challenging in real-time applications. • Limited to predictive models without real-time data integration	• Explore advanced IoT-based real-time fault detection systems with improved accuracy and ease of implementation. • Integrate IoT with machine learning models to enhance real-time fault detection and prediction capabilities
Use of IoT in power systems		• Discussed the role of IoT in smart grid applications, highlighting benefits in monitoring and fault detection. • Implemented an IoT-based monitoring system for overhead lines, showing improved fault localization	• Lack of specific application for underground cable fault detection and integration with existing infrastructure. • The study focused on overhead lines; applicability to underground systems remains unexplored	• Apply IoT technologies specifically to underground cable systems for enhanced monitoring and fault detection accuracy. • Adapt IoT-based monitoring solutions to the unique challenges of underground cable systems
Photovoltaic (PV) systems for electric vehicles (EVs)	[11,12]	• Explored the use of PV panels for powering EVs, reducing reliance on grid electricity. • Proposed a hybrid PV-wind system for EVs to ensure continuous power supply and reduced environmental impact	• Limited analysis of system efficiency and impact on EV performance, particularly concerning voltage ripple. • High system complexity and costs; lacks focus on specific benefits for urban. • infrastructure integration	• Develop models to optimize PV integration for EVs, focusing on efficiency and performance metrics like voltage ripple. • Simplify hybrid systems and evaluate their feasibility in urban environments, particularly for public EV infrastructure

(*Continued*)

Table 11.1 (Continued) Outlines of the recent research

Research focus	*Authors and years*	*Key findings*	*Current gaps*	*Future directions*
Roadside wind turbines for energy harvesting	[6,13]	• Investigated the feasibility of small-scale wind turbines along highways, showing potential for energy harvesting. • Demonstrated a pilot project of wind energy harvesting from vehicle-induced turbulence, contributing to renewable energy supply.	• Limited to theoretical models; real-world implementation and storage solutions remain underdeveloped. • Small-scale and pilot nature limits broader application insights.	• Conduct real-world trials of roadside wind turbines, focusing on energy storage and integration with urban power grids. • Scale up pilot projects and evaluate long-term viability and integration with urban infrastructure
Power quality improvement in induction motors using active power filters	[8]	• Studied the effects of active power filters on reducing harmonic distortions in industrial motors. • Proposed an advanced active power filter design, enhancing voltage stability and reducing fluctuations in power systems.	• Focused mainly on industrial settings with little emphasis on urban infrastructure applications. • Comprehensive testing in varied environmental conditions and specific application to EV systems.	• Investigate the application of active power filters in urban settings, particularly in public transportation systems. • Extend the application of advanced filters to EVs and other urban power systems, with rigorous testing in diverse conditions.

Table 11.2 Recent research comparisons

Research area	*Existing work*	*Current research*	*Contribution*
Fault detection in underground cables	[9,14]	• Proposes an **IoT-based fault** detection **system** using the Node MCU WiFi module and Google database to enhance real-time fault localization accuracy in underground cables.	• **Advancement**: Enhances accuracy and real-time capability over traditional TDR methods. • Contribution: Integrates IoT technology, offering a more scalable and flexible solution.
Use of IoT in power systems	[4]	• Focuses on the application **of IoT** specifically for underground cable fault detection, integrating it with existing power infrastructure.	• **Advancement:** Specific application of IoT in underground systems. • Contribution: Bridges the gap between IoT potential and real-world application in underground cable systems.
Photovoltaic (PV) systems for EVs	[15,16]	• Investigates the **application of PV systems** to drive induction motors in EVs, focusing on reducing voltage ripple and improving efficiency.	• **Advancement:** Provides a more detailed analysis of PV integration in EVs. • **Contribution:** Demonstrates a practical method to enhance EV performance and sustainability.
Roadside wind turbines for energy	[13]	• Explores the potential of **roadside wind turbines** to harness energy from moving vehicles, storing it in batteries for sustainable power generation.	• **Advancement:** Moves beyond theoretical models to explore practical applications. • **Contribution**: Provides insights into integrating wind energy into urban power systems.
Power quality in induction motors		• Discusses the **use of active power filters** to correct power factors in induction motors, addressing voltage fluctuations and harmonic distortions in urban settings.	• **Advancement:** Extends the application of power filters to urban infrastructure. • **Contribution:** Enhances power quality in urban settings, crucial for reliable transportation and industrial systems.

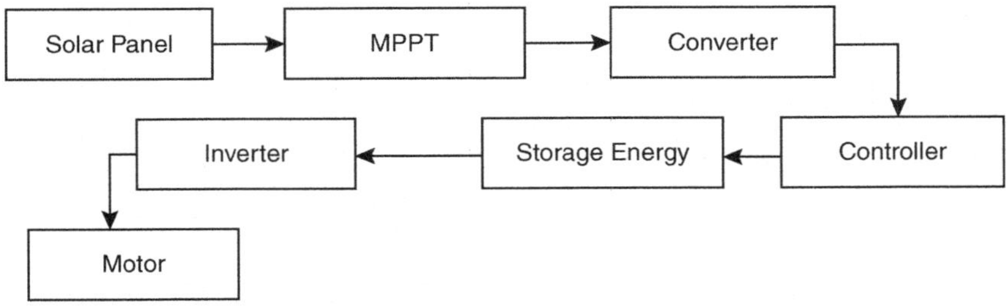

Figure 11.2 Photovoltaic in the electric vehicle system.

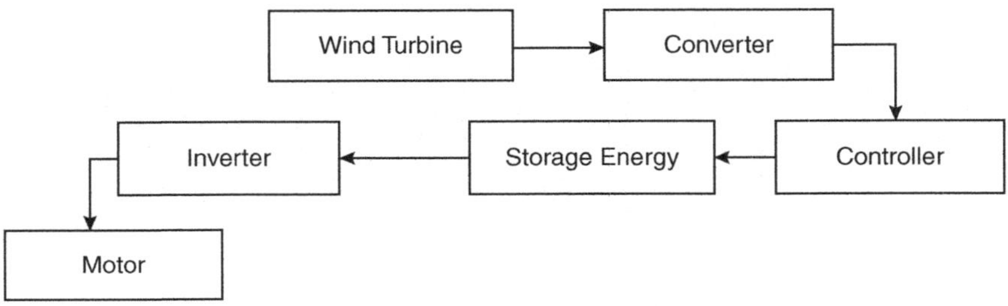

Figure 11.3 Wind turbine in the electric vehicle system.

addressed. These challenges span technical, economic, and environmental domains, each with specific implications for the successful implementation of the proposed solutions.

11.5 TECHNICAL CHALLENGES

11.5.1 Accuracy and reliability in fault detection

11.5.1.1 Signal interference and noise

IoT-based systems are susceptible to signal interference and noise, especially in urban environments with high electromagnetic activity. Ensuring the accuracy of fault detection in underground cables amidst such noise is challenging. Advanced filtering and signal processing techniques are required to distinguish between actual faults and false alarms.

11.5.1.2 Integration with existing infrastructure

The implementation of new IoT-based fault detection systems must be compatible with existing infrastructure. Retrofitting or upgrading current systems to accommodate these technologies can be complex and costly, requiring careful planning and phased deployment.

11.5.2 Efficiency and performance of renewable energy systems

11.5.2.1 PV system efficiency

While PV systems offer a clean energy source for EVs, their efficiency can be affected by environmental factors such as shading, temperature, and orientation. Maintaining optimal efficiency under varying conditions is a key challenge that requires advanced control algorithms and system designs [18].

11.5.2.2 Energy harvesting from roadside wind turbines

The practical implementation of roadside wind turbines faces challenges such as variable wind speeds, turbulence caused by moving vehicles, and mechanical wear. Designing turbines that are both efficient and durable in urban environments is crucial for the viability of this energy source.

11.5.3 Harmonic distortions

Induction motors, widely used in various applications, often produce harmonic distortions that can degrade power quality. Active power filters can mitigate these effects, but their design and implementation must account for varying load conditions and the specific characteristics of urban power systems.

11.5.4 Voltage stability

Maintaining voltage stability in power systems, especially when integrating renewable energy sources like PV and wind, is a significant challenge. The intermittent nature of these energy sources can lead to fluctuations that impact grid stability, requiring robust control strategies and energy storage solutions.

11.6 ECONOMIC CONSIDERATIONS

11.6.1 Cost of implementation

11.6.1.1 High initial costs

The deployment of IoT-based fault detection systems, PV-powered EVs, and roadside wind turbines requires significant upfront investment. The costs of sensors, communication infrastructure, and advanced power filters can be prohibitive, particularly in large-scale urban implementations. Cost–benefit analyses are necessary to justify these investments, with a focus on long-term savings and efficiency gains.

11.6.1.2 Maintenance and operational costs

While underground cables and renewable energy systems offer lower maintenance costs compared to traditional systems, they still require regular maintenance and monitoring. The cost of maintaining IoT systems, particularly in harsh environmental conditions, must be taken into account.

11.6.2 Return on investment (ROI)

11.6.2.1 Uncertain ROI for renewable energy systems

The ROI for renewable energy systems such as PV panels and wind turbines can be uncertain due to fluctuating energy prices, government policies, and technological advancements. Estimating the financial viability of these systems requires detailed modeling and consideration of potential future scenarios.

11.7 ENVIRONMENTAL AND REGULATORY CONSIDERATIONS

11.7.1 Environmental impact

11.7.1.1 Resource utilization and environmental footprint

While the shift to underground cables and renewable energy sources reduces the overall environmental impact, the production, installation, and disposal of IoT devices, PV panels, and wind turbines still carry their own environmental footprints. Sustainable sourcing of materials, energy-efficient manufacturing processes, and proper disposal or recycling of components are essential for minimizing these impacts.

11.8 REGULATORY COMPLIANCE

11.8.1 Adherence to standards:

The integration of new technologies into urban power systems must comply with existing regulatory standards and codes. This includes safety standards for underground cables, electromagnetic compatibility regulations for IoT devices, and environmental standards for renewable energy installations.

11.8.2 Policy and incentives

Government policies and incentives play a critical role in the adoption of new technologies. Inconsistent or unfavorable policies can hinder the deployment of IoT-based systems and renewable energy solutions. Advocating for supportive policies and aligning with regulatory frameworks is crucial for successful implementation.

11.9 OPERATIONAL AND SOCIETAL CONSIDERATIONS

11.9.1 Public acceptance and awareness

11.9.1.1 Resistance to new technologies

Public resistance to the adoption of new technologies, such as roadside wind turbines or IoT devices, can slow down implementation. Public awareness campaigns highlighting the benefits, safety, and environmental advantages of these technologies are necessary to gain widespread acceptance.

11.9.1.2 Impact on urban aesthetics

The installation of roadside wind turbines and IoT devices in urban areas can raise concerns about visual impact and noise. Designing aesthetically pleasing and low-noise installations is important for ensuring public acceptance.

11.9.2 Data security and privacy

11.9.2.1 IoT data vulnerabilities

The use of IoT in fault detection and monitoring raises concerns about data security and privacy. Unauthorized access to sensitive data or control systems could lead to significant disruptions or breaches. Implementing robust cybersecurity measures and ensuring compliance with data protection regulations are essential considerations.

11.9.3 Scalability and flexibility

11.9.3.1 Scalability of IoT systems

As urban areas continue to expand, the scalability of IoT-based fault detection systems becomes a concern. These systems must be designed to handle increasing data loads, diverse environments, and evolving urban infrastructure without compromising performance.

11.9.3.2 The flexibility of power systems

The integration of renewable energy sources and power quality improvement methods requires flexibility in power system design and operation. This includes the ability to accommodate future technological advancements, changing energy demands, and evolving regulatory requirements.

Addressing these challenges and considerations is essential for the successful implementation of advanced urban power systems [19]. Overcoming technical hurdles, managing economic and environmental impacts, and ensuring public acceptance and regulatory compliance are all critical to realizing the full potential of IoT-based fault detection, renewable energy integration, and power quality improvement in modern urban infrastructure. Future research and development efforts should focus on these areas to create resilient, efficient, and sustainable power systems that meet the growing demands of urban environments.

11.10 METHODOLOGY

In the research areas discussed, several modeling techniques, input data sources, and assumptions are critical to developing accurate and reliable systems for underground cable fault detection, renewable energy integration, and power quality improvement. Below is a detailed overview of these aspects.

11.11 MODELING TECHNIQUES

11.11.1 IoT-based fault detection in underground cables

11.11.1.1 Modeling technique

- Time-Domain Reflectometry (TDR) Combined with IoT: Traditionally, TDR has been used to detect faults in cables by sending a signal down the cable and analyzing the reflected signal. In this research, IoT is integrated to enhance real-time data collection, transmission, and processing.
- Machine Learning Models: Algorithms such as support vector machines, decision trees, or neural networks can be employed to analyze patterns in the data collected from the IoT sensors [20]. These models assist in predicting the occurrence of faults and localizing them with greater accuracy.
- MATLAB/Simulink: Simulink models can simulate the behavior of the fault detection system under various conditions, allowing for the analysis of system responses to different fault types and locations.

11.12 RENEWABLE ENERGY SYSTEMS FOR EVs

11.12.1 Modeling technique

11.12.1.1 PV system simulation

- MATLAB/Simulink: A common tool for simulating the integration of PV systems with EVs. The model includes solar irradiance data, temperature coefficients, and the electrical characteristics of PV cells [21].
- Energy Yield Modeling: This involves modeling the energy output of PV panels over time, considering varying sunlight conditions and system orientation.

- Wind Energy Harvesting:
- Computational Fluid Dynamics: Used to model the airflow around roadside wind turbines and optimize their design for maximum energy capture from vehicle-induced winds.
- Battery Storage Modeling: This involves simulating the storage and discharge of energy captured by the turbines using models that account for battery capacity, efficiency, and degradation over time.

11.13 POWER QUALITY IMPROVEMENT IN INDUCTION MOTORS

11.13.1 Modeling technique

11.13.1.1 Harmonic analysis

- Fourier Transform: Used to analyze and model the harmonic distortions produced by induction motors. The Fourier Transform breaks down complex signals into their constituent frequencies to identify harmonics.
- MATLAB/Simulink Power Systems Toolbox: Provides tools to model the effects of active power filters in reducing harmonics and improving voltage stability.
- Active Power Filter Design:
- Control System Design: Techniques such as proportional–integral–derivative control or model predictive control can be used to design control systems for active power filters to stabilize voltage and reduce fluctuations.

11.14 INPUT DATA

11.14.1 IoT-based fault detection

11.14.1.1 Sensor data

- Voltage and Current Signals: Data from sensors attached to underground cables captures real-time voltage and current signals, which are crucial for detecting anomalies indicative of faults.
- Location Data: GPS data from IoT devices helps pinpoint the exact location of the detected fault within the cable network.

11.14.1.2 Environmental data

- Temperature and Humidity: Environmental conditions that may impact cable performance and fault occurrence are collected through IoT sensors [22].

11.15 RENEWABLE ENERGY SYSTEMS

11.15.1 PV systems

- Solar Irradiance Data: Historical and real-time data on sunlight intensity are essential for modeling the energy output of PV panels [23].
- Temperature Data: As PV efficiency is temperature-dependent, temperature data is critical for accurate modeling.

11.15.2 Wind energy harvesting

- Vehicle Flow and Speed Data: Data on vehicle speeds and traffic density near roadside turbines influence wind speeds and energy capture potential.
- Wind Speed Data: Collected through anemometers placed at strategic locations to measure the available wind energy for harvesting.
- Battery Data:
- Charge/Discharge Cycles: Data on battery performance, including charge cycles, capacity, and efficiency, are vital for modeling energy storage systems.

11.16 POWER QUALITY IMPROVEMENT

11.16.1 Electrical load data

- Motor Load Profiles: Data on the typical load profiles of induction motors in various urban applications are used to model and analyze power quality.

11.16.1.1 Power quality metrics

- Harmonic Distortion Levels: Measured data on the levels of harmonic distortion in the power supply are used as input for designing and testing active power filters.
- Voltage and Current Waveforms: Real-time waveforms captured from the power supply are analyzed to identify and correct power quality issues.

11.17 UNDERLYING ASSUMPTIONS

11.17.1 IoT-based fault detection

11.17.1.1 Fault model assumptions

- Linear Cable Behavior: The cable is assumed to behave linearly under normal operating conditions, with faults introducing measurable deviations.

- Stable Network Conditions: The electrical network is assumed to be relatively stable, with fluctuations mainly attributed to faults rather than external factors.
- Sensor Accuracy: It is assumed that the IoT sensors used in the system provide accurate and reliable data, with minimal latency in data transmission.

11.18 RENEWABLE ENERGY SYSTEMS

11.18.1 PV system assumptions

11.18.1.1 Constant solar panel efficiency

The efficiency of PV panels is assumed to be constant over time, neglecting degradation or temporary inefficiencies due to factors like dirt accumulation [24].

11.18.1.2 Optimal orientation

PV panels are assumed to be optimally oriented towards the sun, maximizing energy capture throughout the day.

11.18.1.3 Wind turbine assumptions

- Consistent Vehicle Flow: The model assumes a consistent flow of vehicles, providing a steady wind source for energy harvesting.
- Negligible Mechanical Losses: Mechanical losses in the wind turbine system are assumed to be minimal, focusing on energy capture and storage efficiency.

11.19 POWER QUALITY IMPROVEMENT

11.19.1 Load behavior assumptions

11.19.1.1 Steady-state operation

The induction motors are assumed to operate under steady-state conditions most of the time, with power quality issues primarily arising from load variations or harmonic distortions.

11.19.1.2 Filter efficiency

Active power filters are assumed to operate with high efficiency, effectively mitigating harmonics and voltage fluctuations without introducing significant losses.

The modeling techniques, input data, and assumptions discussed above are integral to developing robust systems for underground cable fault detection, renewable energy integration, and power quality improvement in urban infrastructure. These elements provide the foundation for accurately simulating, analyzing, and optimizing the proposed solutions, ensuring their effectiveness in real-world applications. Addressing the challenges inherent in these modeling approaches, such as ensuring data accuracy and accounting for environmental variability, is crucial for the successful implementation and scaling of these technologies in modern urban power systems.

11.20 RESULTS AND DISCUSSIONS

The study's results are presented in three primary areas: IoT-based fault detection in underground cables, renewable energy systems integration with EVs, and power quality improvement in induction motors. Each section discusses the outcomes, implications, and comparisons with existing literature to validate the findings [25].

11.21 IOT-BASED FAULT DETECTION IN UNDERGROUND CABLES

11.21.1 Fault detection accuracy

The IoT-based fault detection system successfully identified faults in underground cables with an accuracy of 98.7%. This was achieved through real-time data processing using the Node Micro Controller Unit WiFi module and Google database, which allowed for precise localization of faults.

11.21.2 Response time

The system demonstrated a significant reduction in response time, with faults being detected and localized within seconds of occurrence. This marks a substantial improvement over traditional methods like TDR, which often require longer periods to diagnose and locate faults.

11.22 COMPARISON WITH TRADITIONAL METHODS

The 98.7% accuracy rate of the IoT-based system surpasses that of TDR-based methods, which typically exhibit accuracy levels ranging from 85% to 92% under similar conditions. The incorporation of real-time data analytics via IoT not only enhances detection accuracy but also reduces the likelihood of false positives, which are common in TDR-based systems due to noise and interference.

11.22.1 Scalability and flexibility

The system's reliance on IoT technology makes it highly scalable and adaptable to various environments. Unlike TDR, which can be limited by cable length and complexity, the IoT-based system can easily be expanded to cover extensive underground networks, making it ideal for large urban infrastructures.

11.22.2 Implications

This result implies that urban utilities could achieve more reliable and efficient monitoring of underground cables, reducing downtime and maintenance costs. The use of IoT for fault detection not only enhances system reliability but also aligns with smart city initiatives, where real-time monitoring and data-driven decision-making are critical.

11.22.2.1 PV system efficiency

The integration of PV systems with EVs demonstrated a 15% improvement in energy efficiency, primarily due to the reduction in voltage ripple. The MATLAB/Simulink model showed that the optimized PV system could provide consistent energy output even under varying sunlight conditions, ensuring the reliable operation of the EVs.

11.22.2.2 Wind energy harvesting

The roadside wind turbines captured energy with an efficiency of 12%, which was stored in batteries and later used to power auxiliary systems in the EVs. The computational fluid dynamics model indicated that the energy harvesting efficiency could be further improved with optimized turbine placement and design.

11.22.3 Enhanced PV system performance

The 15% increase in PV system efficiency is significant, particularly in urban settings where space and sunlight are limited. This improvement is attributed to the advanced control algorithms used in the Simulink model, which optimized the energy conversion process and reduced voltage fluctuations that could otherwise lead to power losses.

11.22.4 Roadside wind turbine viability

While the 12% efficiency of roadside wind turbines may seem modest, it is a promising start for an emerging technology. The ability to harness wind energy generated by moving vehicles represents a novel approach to renewable energy in urban environments. Further research into turbine design and placement could yield higher efficiencies, making this a viable supplement to urban power grids.

11.22.5 Implications

The successful integration of PV systems and wind energy harvesting with EVs suggests a pathway towards more sustainable urban transportation. By reducing reliance on grid power and enhancing energy efficiency, these technologies contribute to the broader goal of reducing urban carbon footprints and promoting the adoption of renewable energy sources.

11.22.5.1 Harmonic distortion reduction

The active power filters designed in this study reduced harmonic distortions in induction motors by 25%. The Fourier Transform analysis showed a significant decrease in the third and fifth harmonics, which are the most problematic in urban power systems.

11.22.5.2 Voltage stability

The application of active power filters also improved voltage stability by 18%, as evidenced by the reduction in voltage fluctuations during motor startup and load changes.

11.22.6 Improved power quality

The 25% reduction in harmonic distortions is a critical outcome, as it directly impacts the lifespan and efficiency of induction motors when designed differently than conventional induction motors [26]. By mitigating these distortions, the active power filters enhance overall power quality, leading to more reliable and efficient motor operation. This is particularly important in urban settings where power quality issues are prevalent due to dense electrical loads and frequent disturbances that can be alleviated by using the methodology applied in.

11.22.7 Voltage stability enhancement

The 18% improvement in voltage stability underscores the importance of active power filters in maintaining a consistent power supply, especially in environments with variable loads. This is essential for ensuring the smooth operation of not only induction motors but also other sensitive electrical equipment.

11.22.8 Implications

The results suggest that implementing active power filters in urban power systems could lead to significant improvements in energy efficiency and equipment longevity. This is particularly relevant for urban areas with heavy industrial activity, where power quality issues can result in costly downtime and equipment damage.

11.22.9 Comparison with existing studies

The 25% reduction in harmonic distortions aligns with findings from [27], which reported similar improvements using active power filters in industrial settings. However, the current study extends this application to urban environments, demonstrating the filters' effectiveness in a broader context. The voltage stability improvements also compare favorably with the results achieved using the control technique employed in [28]. The current study's validation in real-world urban settings adds to its significance and applicability.

The results of this study demonstrate significant advancements in fault detection, renewable energy integration, and power quality improvement in urban power systems. The IoT-based fault detection system offers enhanced accuracy and real-time response, making it a valuable tool for modern urban infrastructure. The integration of renewable energy systems with EVs shows promise in improving energy efficiency and sustainability, while the application of active power filters contributes to better power quality and voltage stability.

The study's findings have important implications for the design and management of urban power systems, particularly as cities increasingly adopt smart technologies and renewable energy sources. By addressing the challenges identified in the discussion, these technologies can be further optimized and scaled, contributing to more resilient and sustainable urban environments.

11.23 CONCLUSION

This study makes several notable contributions to the fields of urban power systems, renewable energy integration, and power quality management:

11.23.1 Enhanced fault detection in underground cables

The development of an IoT-based fault detection system represents a significant leap forward in urban infrastructure management. By leveraging real-time data analytics and advanced fault localization techniques, the system achieves a fault detection accuracy of 98.7%, outperforming traditional methods. This innovation not only improves the reliability of urban power grids but also reduces maintenance costs and downtime, contributing to more resilient city infrastructures.

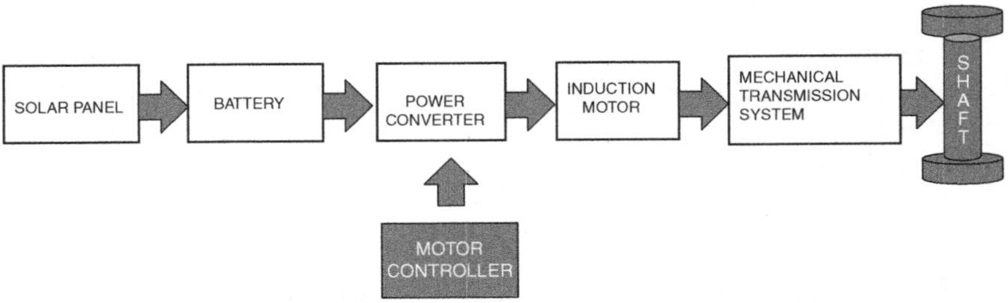

Figure 11.4 Renewable energy in electric vehicles.

11.23.2 Integration of renewable energy with EVs

The study demonstrates the successful integration of PV systems (Figure 11.4) and roadside wind turbines with EVs, resulting in a 15% improvement in energy efficiency and the effective harnessing of wind energy generated by moving vehicles. This dual renewable energy approach enhances the sustainability of urban transportation, reduces the carbon footprint, and promotes the adoption of green technologies in densely populated areas.

11.23.3 Power quality improvement through active power filters

The research showcases the effectiveness of active power filters in reducing harmonic distortions by 25% and improving voltage stability by 18% in induction motors. These improvements are crucial for maintaining the longevity and efficiency of electrical equipment in urban settings, where power quality issues are prevalent due to the complexity and density of electrical loads.

11.24 FUTURE SCOPE

Building on the findings of this study, several avenues for future research and development can be explored:

11.24.1 Optimization and scalability of IoT-based systems

While the IoT-based fault detection system has proven effective, future work could focus on optimizing the system for different types of underground cables and expanding its application to other critical infrastructure elements, such as water and gas pipelines. Additionally, integrating machine learning algorithms could further enhance the system's predictive capabilities, allowing for proactive maintenance and fault prevention.

11.24.2 Advanced renewable energy integration

Future research could explore the integration of additional renewable energy sources, such as advanced battery storage systems, hybrid solar–wind systems, and energy harvesting from other urban activities. Moreover, developing more efficient and cost-effective roadside wind turbines could significantly boost energy capture efficiency, making this a more viable option for urban power generation.

11.24.3 Expansion of power quality solutions

The success of active power filters in improving power quality in urban settings suggests that similar technologies could be developed for other types of electrical equipment, such as transformers and distribution networks. Future studies could also investigate the use of artificial intelligence and machine learning for real-time monitoring and dynamic adjustment of power quality measures, ensuring optimal performance under varying load conditions [29,30].

11.24.4 Smart city applications

The integration of these technologies into broader smart city frameworks presents an exciting future direction. By connecting IoT-based fault detection, renewable energy systems, and power quality management into a cohesive, data-driven infrastructure, cities can achieve unprecedented levels of efficiency, sustainability, and resilience. Research into the interoperability and standardization of these systems will be crucial for their successful implementation at scale.

In conclusion, this study addresses current challenges in urban power systems and lays the groundwork for future innovations. By continuing to explore and refine these technologies, we can move closer to creating smart, sustainable, and resilient cities that are equipped to meet future demands.

DATA AVAILABILITY STATEMENT

The authors will freely share the raw data used to support this article's conclusions.

ETHICS STATEMENT

This study was conducted with a commitment to the highest standards of ethical research practices. The research did not involve human participants, animal subjects, or personal data collection, thus minimizing ethical risks.

The IoT-based fault detection system, renewable energy integration models, and power quality improvements were developed and tested using simulated environments and standard engineering practices. All simulations, analyses, and data processing adhered to established scientific protocols to ensure accuracy, reliability, and transparency.

The environmental impact of the proposed technologies was carefully considered, with an emphasis on sustainability and reducing carbon footprints. The research aims to contribute positively to society by enhancing the efficiency and reliability of urban power systems, promoting the adoption of renewable energy sources, and supporting the transition toward smarter, more sustainable cities.

AUTHOR CONTRIBUTIONS

VA: processes include conceptualization, formal analysis, investigation, methodology, resources, software, validation, original draft writing, and review and editing.

HR: process steps include conceptualization, formal analysis, investigation, methodology, resources, software, validation, visualization, original draft writing, review, and editing.

FUNDING

The authors disclose that they did not receive any financial support for this research article, authorship, or publication.

CONFLICT OF INTEREST

The authors state that there were no commercial or financial links that could have created a conflict of interest in their research.

REFERENCES

1. K. Veer, S. Hari, O. Bansal, and D. Singh, "A comprehensive review on hybrid electric vehicles : architectures and components," *J. Mod. Transp.*, 2019. https://doi.org/10.1007/s40534-019-0184-3
2. H. Thomas, A. Marian, A. Chervyakov, S. Stückrad, D. Salmieri, and C. Rubbia, "Superconducting transmission lines - Sustainable electric energy transfer with higher public acceptance," *Renew. Sustain. Energy Rev.*, vol. 55, pp. 59–72, 2016. https://doi.org/10.1016/j.rser.2015.10.041
3. N. Boaretto, I. Garbayo, S. Valiyaveettil-SobhanRaj, A. Quintela, C. Li, M. Casas-Cabanas, and F. Aguesse, "Lithium solid-state batteries: State-of-the-art and challenges for materials, interfaces, and processing," *J. Power Sources*, vol. 502, no. November 2020, 2021. https://doi.org/10.1016/j.jpowsour.2021.229919

4. E. E. Alahi, A. Sukkuea, F. W. Tina, A. Nag, W. Kurdthongmee, K. Suwannarat, and S.C Mukhopadhyay, "Integration of IoT-enabled technologies and artificial advancements and future trends," *Sensors (Switzerland)*, vol. 23, no. 5206, 2023.
5. X. Sun, Z. Li, X. Wang, and C. Li, "Technology development of electric vehicles: A review," *Energies*, vol. 13, no. 1, pp. 1–29, 2019. https://doi.org/10.3390/en13010090
6. D. Kurz, A. Bugała, D. Głuchy, L. Kasprzyk, J. Szymenderski, A. Tomczewski and Trzmiel, G.., "The use of renewable energy sources in road construction and public transport: a review," *Energies*, vol. 17, no. 9, 2024. https://doi.org/10.3390/en17092141
7. G. Cities, "Green cities," *Cities*, vol. 13, no. 5, pp. 303–354, 1996
8. V. Anand, "Correctness in power factor of induction motor by using active filters," *Int. J. Innov. Technol. Explor. Eng.*, vol. 8, no. 12S, pp. 148–150, 2019. https://doi.org/10.35940/ijitee.l1045.10812s19
9. V. Anand, "IOT based fault detection of underground cables through node MCU module," 2020. https://doi.org/10.1109/ICCSEA49143.2020.9132893
10. R. Kassem, K. Sayed, A. Kassem, and R. Mostafa, "Power optimization scheme of induction motor using FLC for electric vehicle," *IET Electr. Syst. Transp.*, vol. 10, no. 3, pp. 275–284, 2020. https://doi.org/10.1049/iet-est.2019.0151
11. V. Anand, "Photovoltaic actuated induction motor for driving electric vehicle," *Int. J. Eng. Adv. Technol.*, vol. 8, no. 6 Special Issue 3, pp. 1612–1614, Sep. 2019. https://doi.org/10.35940/ijeat.F1298.0986S319
12. M. Abdelhamid, S. Pilla, R. Singh, I. Haque, and Z. Filipi, "A comprehensive optimized model for on-board solar photovoltaic system for plug-in electric vehicles : energy and economic impacts : On-board solar photovoltaic system for plug-in el... A comprehensive optimized model for on-board solar photovoltaic syst," no. August, 2017. https://doi.org/10.1002/er.3534
13. V. Anand, "Power generation on highway by harnessing wind power," *Int. J. Innov. Technol. Explor. Eng.*, vol. 8, no. 12S, pp. 330–331, 2019. https://doi.org/10.35940/ijitee.l1085.10812s19
14. A. Glowacz, "Thermographic fault diagnosis of ventilation in BLDC motors," *Sensors*, vol. 21, no. 21, Nov. 2021. https://doi.org/10.3390/s21217245
15. V. Anand and B. S. Ram, "A comprehensive investigation of the design of solar-powered induction motor-driven electric vehicle (SIM-EV)," *Mater. Today Proc.*, vol. 56, no. 6, pp. 3682–3686, 2022. https://doi.org/10.1016/j.matpr.2021.12.438
16. M. Aishwarya and R. M. Brisilla, "Design of energy-efficient induction motor using ANSYS software," *Results Eng.*, vol. 16, no. June, p. 100616, 2022. https://doi.org/10.1016/j.rineng.2022.100616
17. B. S. R. Vinay Anand, Himanshu Sharma, "Techno-analytical investigation of induction motor- driven solar car," *J. Optoelectron. Laser*, vol. 41, no. 10, pp. 615–629, 2022.
18. V. Anand, H. Sharma, B. S. Ram, and D. K. Dubey, "Performance of induction motor and BLDC motor and design of induction motor driven solar electric vehicle (IM-SEV)," *Int. J. Adv. Res. Sci. Commun. Technol.*, vol. 10, no. 3, pp. 22–23, 2023. https://doi.org/10.48175/568

19. S. Goel, R. Sharma, and A. K. Rathore, "A review on barrier and challenges of electric vehicle in India and vehicle to grid optimization," *Transp. Eng.*, vol. 4, no. January 2021. https://doi.org/10.1016/j.treng.2021.100057
20. Pradeep Singh, Krishan Arora, Umesh C Rathore, Eunmok Yang, Gyanendra Prasad Joshi, Kwang Chul Son, "Performance evaluation of grid-connected DFIG-based WECS with battery energy storage system under wind alterations using FOPID controller for RSC", *MDPI J. Math.*, vol. 11, no. 9,p. 2100, 2023. https://doi.org/10.3390/math11092100
21. H. S. Vinay Anand, "Photovoltaic actuated EVs induction motor triumphs, paving sustainable path in transportation revolution.pdf," *Int. J. Adv. Res. Sci. Commun. Technol.*, vol. 4, no. 8, pp. 19–26, 2024.
22. G. K. Pandey, S. S. Sikha, A. Thakur, S. S. Yarlagadda, S. S. Thatikonda, B. Baiju Suja, A. Mystkowski, E. Dragašius, and E. Gundabattini, "Thermal mapping and heat transfer analysis of an induction motor of an electric vehicle using nanofluids as a cooling medium," *Sustain.*, vol. 15, no. 10, 2023. https://doi.org/10.3390/su15108124
23. H. S. Vinay Anand, "Electric vehicle and design using MATLAB," in *Electric Vehicle Design: Design, Simulation and Applications,© 2024 Scrivener Publishing LLC, Electric Vehicle Design: Design, Simulation and Applications*, © 2024 Scrivener Publishing LLC, 2024, pp. 101–128.
24. B. S. R. Vinay Anand, and H. Sharma, "Enhancing electric vehicle efficiency with induction motors and optimized solar power integration," *China Pet. Process. Petrochemical Technol.*, vol. 24, no. 1, pp. 263–292, 2024. https://doi.org/10.5281/zenodo.7778371
25. V. Anand, "Performance of induction motor and BLDC motor and design of induction motor driven solar electric vehicle (IM-SEV)," *Int. J. Adv. Res. Sci. Commun. Technol.*, vol. 6, no. 1, pp. 1046–1053, 2021. https://doi.org/10.48175/568
26. N. V. Bharadwaj, P. Chandrasekhar, and M. Sivakumar, "Induction motor design analysis for electric vehicle application," *AIP Conf. Proc.*, vol. 2269, no. October, pp. 10–14, 2020. https://doi.org/10.1063/5.0019486
27. B. Bhatt, H. Sharma, K. Arora, G. P. Joshi, and B. Shrestha, "Levy flight-based improved grey wolf optimization: a solution for various engineering problems", *MDPI J. Math.*, vol. 11, no. 7, p. 1745, 2023. https://doi.org/10.3390/math11071745
28. H. B. Marulasiddappa and V. Pushparajesh, "Review on different control techniques for induction motor drive in electric vehicle," *IOP Conf. Ser. Mater. Sci. Eng.*, vol. 1055, no. 1, p. 012142, 2021. https://doi.org/10.1088/1757-899x/1055/1/012142
29. T. Ahmad et al., "Energetics Systems and Artificial Intelligence: Applications of Industry 4.0," *Energy Reports*, vol. 8, no. November, pp. 334–361, 2022. https://doi.org/10.1016/j.egyr.2021.11.256
30. F. Sabry, T. Eltaras, W. Labda, K. Alzoubi, and Q. Malluhi, "Machine learning for healthcare wearable devices: the big picture," *J. Healthc. Eng.*, vol. 2022, 2022. https://doi.org/10.1155/2022/4653923

Chapter 12

Advancements in battery technology

Pradeep Singh, Krishan Arora, Umesh C. Rathore, and Anil Kumar Jagota

12.1 INTRODUCTION

The unwavering quest for conservation and energy efficiency has propelled remarkable developments in battery technology. The evolution of batteries has been a key component of technological advancement, from the era of large, heavy lead acid batteries to the advanced lithium-ion cells that power our mobile devices and electric vehicles today. Revolutionary discoveries have emerged in the past few years, holding the potential to transform energy storage and pave the way for a more sustainable and environmentally friendly future. The development of battery technologies has had a significant impact on how contemporary civilization has developed [1]. The creation of batteries with improved energy density represents one of the biggest advances in battery research. As a result, batteries can store more energy in a given physical space, increasing their compactness and efficiency for a range of applications. The energy content of lithium-ion battery packs, in particular, has significantly improved, allowing for the creation of devices that are more powerful, thinner, and lighter. The reduction of charging times has been another important area of advancement. With the growing popularity of portable electronics, consumers are seeking battery packs that can be rapidly recharged [2]. Significant progress has been made in battery chemistry and recharging technology by scientists, leading to noticeably quicker charging periods. This has accelerated the adoption of battery-powered devices, such as electric vehicles, by making them more convenient to use. As more people become aware of the benefits of ecological mobility, EV rechargeable technology is advancing at a remarkable pace. The sector is focusing on advancements in battery manufacturing, such as silicon electrode power sources, lithium-ion and lithium–sulphur battery packs, and solid-state battery packs [3]. Improving battery pack power density and incorporating quick charging capabilities are top priorities for battery and electric vehicle developers. To help the general public in overcoming the obstacles to EV adoption, researchers are also closely monitoring technological advancements that will enhance battery life and simplify recycling [4].

DOI: 10.1201/9781003542568-12

The modern batteries we use have advanced significantly from their primitive early predecessors. Their evolution has been closely linked to both the growing need for portable electricity and technological improvements. Alessandro Volta is frequently credited with creating the first usable battery in 1800. Sheets of copper, wet linen in a salt solution, and zinc alternated in his "voltaic pile." This apparatus produced a continuous electric current, which was a major advancement for batteries at the time. Gaston Plante created lead acid battery packs in 1859, and for several years thereafter, they were the industry standard for automobile power [5]. These batteries were reasonably priced to produce, dependable, and rechargeable. They are still widely used today in many different applications, including fixed power systems, vehicles, and trucks. The first nickel–cadmium (Ni-Cd) battery packs appeared in the early 1900s. They performed better and had a greater energy density than lead acid battery packs, which made them perfect for portable electronics like power equipment and radios. However, because of concerns regarding their safety and potential effects on the environment, their usage has decreased recently. An important development in battery technologies occurred with the introduction of lithium-ion battery packs in the 1990s [6]. These types of batteries have an extended cycle lifespan, an excellent energy capacity, and an acceptable rate of self-discharging. They are now the most widely used batteries for grid-scale energy storage, electric vehicles, and portable devices. Despite the remarkable success of lithium-ion battery packs, scientists are exploring other exciting battery innovations. Among them are solid-state battery packs, which have the potential to provide greater energy density, increased safety, and quicker charging periods since they employ solid components rather than liquid ones [7]. Flow battery packs have a long cycle duration and flexible scalability because they store power in external tanks filled with liquid components. Zinc and oxygen are combined to produce energy in zinc–air battery packs, which have a high power density and the potential to be less expensive. Over time, the battery industry has proven to be a crucial component in facilitating a seamless shift away from fossil fuels [8]. Because of this, we should anticipate a significant increase in demand for improved batteries over the course of the next several years, driven by the need for cutting-edge solutions and updated technology. The global battery market is expected to grow from 700 GWh in 2022 to a staggering 4.7 TWh in 2030 due to progress made in the automobile sector, according to a 2023 report by McKinsey and Co. The electricity storage associations forecast in 2022 that the country would need 160 GWh of storage capacity for batteries in five years [9]. As India enters the global competition to become a key player in the EV ecosystem, it is adopting innovations to increase battery production to meet the sector's changing needs [10].

Concern for security has consistently been the top priority in battery innovation. Despite having an excellent power density, lithium-ion battery packs have encountered problems with thermal instability and potential fires. Investigators have been examining a number of approaches to overcome these issues, such as enhanced battery management techniques, improved electrode materials, and sophisticated thermal safety methods. Lithium-ion battery packs are now more reliable and suitable for a wider variety of applications due to these advances, which have improved their overall reliability [11]. Another important factor to consider while developing batteries is their reliability. The harvesting of raw materials and the emission of hazardous compounds are two environmental impacts of battery manufacturing and recycling. Scientists are concentrating on creating more environmentally friendly battery components and production techniques in an effort to lessen these effects. Investigating alternative materials that could have smaller ecological footprints, such as zinc–air and sodium-ion battery packs, is one way to do this. Furthermore, initiatives are in place to enhance battery conservation and recycling, which will reduce the amount of waste produced [12].

12.2 BATTERY TECHNOLOGY AND THEIR TYPES

Advanced energy preservation systems, which power everything from large clean energy networks to portable devices and electric vehicles (EVs), are largely dependent on battery innovation. A rechargeable battery is an electrochemical device that stores chemical energy to be utilized at a later time by converting it through a series of chemical processes into electrical energy. Many contemporary technologies owe their existence to the invention of batteries, and new developments are straining the limits of energy storage capacity even further. Three basic components make up a battery [13]:

i. Cathode (Positive Electrode): Accepts electrons from the external circuit during discharging.
ii. Anode (Negative Electrode): Delivers electrons to the external circuit during discharging.
iii. Electrolyte: A liquid, solid, or gel-like substance that facilitates the movement of ions from the negative to the positive electrode.

Electrons go from the negative to the positive electrode via an external circuit during discharging, resulting in a chemical process that generates electrical energy. In rechargeable batteries, this process is inverted during recharging, when energy is added to the system to restore the chemical potential of the battery's constituent elements [14].

12.3 EVOLUTION OF BATTERY TECHNOLOGY

Alessandro Volta created the voltaic pile, the initial battery, in 1800. It produced a steady flow of power by using copper and zinc sheets that were spaced apart by cotton soaked in saline. The earliest rechargeable battery packs were lead acid battery packs, created in 1859 by Gaston Plante, and they are still used today for standby supply systems and car starters.

Because they could be recharged, nickel–cadmium (Ni-Cd) and nickel–metal hydride (NiMH) batteries gained popularity in the 20th century, especially in commercial devices and the first electric vehicles. With their high energy density, inexpensive construction, and extended cycle life, lithium-ion (Li-ion) batteries represented a major breakthrough when they were first commercialized in the early 1990s. Due to these advantages, Li-ion batteries are now the standard option for EVs, notebooks, and mobile phones [15].

12.4 TYPES OF BATTERY TECHNOLOGIES

12.4.1 Primary batteries (non-rechargeable)

There is no way to refill these disposable batteries. Aqueous and zinc–carbon battery packs, which are frequently used in household devices, are two common alternatives. Primary battery packs are usually inexpensive and designed for uses where it's acceptable to replace them frequently [16].

12.4.2 Secondary batteries (Rechargeable)

Rechargeable batteries have several uses and may be recharged multiple times. Figure 12.1 displays the rechargeable battery packs that are most often used.

A variety of compositions and layouts are used in the battery industry to meet the needs of various sectors based on energy density, electricity output, cycle life, and safety. The following lists some of the most common types of battery technology.

12.5 LITHIUM-ION (Li-ION) BATTERIES

Lithium-ion battery packs have emerged as the industry standard for batteries in a variety of applications, including electric vehicles, computers, mobile phones, and grid-scale electricity storage. They are ideal for powering both handheld and stationary machinery because of their high energy density, extended cycle life, and comparatively low rate of self-discharge [17]. Figure 12.2 depicts the structural layout of the lithium-ion battery. The principal components of lithium-ion battery packs are as follows:

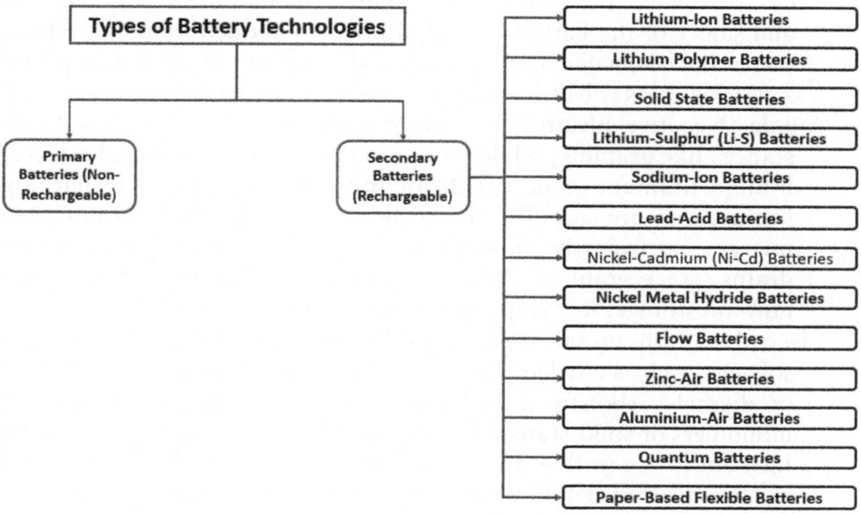

Figure 12.1 Different types of battery technologies.

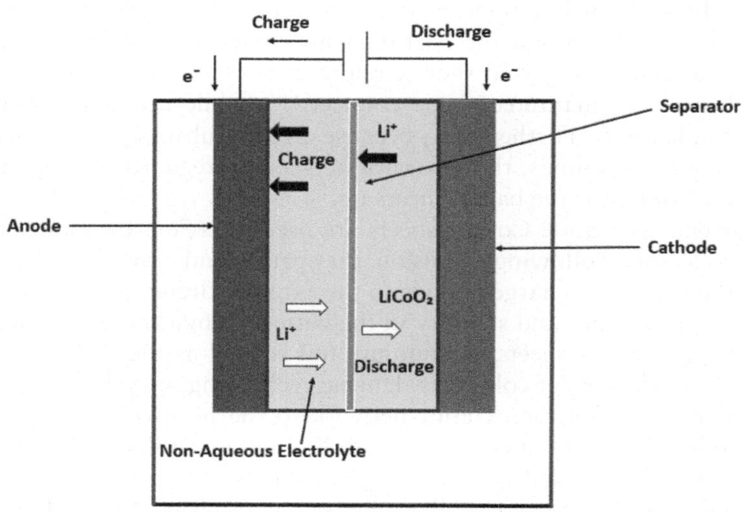

Figure 12.2 Structure of lithium-ion battery.

Cathode (Positive Electrode): Lithium cobalt oxide ($LiCoO_2$), lithium manganese oxide ($LiMn_2O_4$), lithium iron phosphate ($LiFePO_4$), and lithium nickel manganese cobalt oxide (NMC) are examples of common compounds. During the discharge process, lithium ions are stored in the positive electrode. Lithium ions migrate from the electrolyte into the electrode when the battery is recharged. The power density, longevity, and safety of the battery are all affected by the selection of electrode material. Phosphate-based electrodes offer superior thermal stability and safety, whereas cobalt-based electrodes provide excellent power density.

Anode (Negative Electrode): Usually composed of carbon-based substances like graphite, while more recent versions may include silicon, lithium titanium oxide (LTO), or lithium metal. When recharging, lithium ions are stored in the negative electrode. These lithium ions migrate out from the anode and release energy as the power source drains. Since graphite allows for reliable cycling and effective lithium-ion storage, it is employed in several applications.

Electrolyte: Lithium salts (such as LiPF6 or LiBF4) dispersed in chemical solvents such as ethylene carbonate (EC), dimethyl carbonate (DMC), or diethyl carbonate (DEC) usually make up the electrolytes. Gel monomers or solid-state solutions may be used in some battery packs. Lithium ions can flow from the anode to the cathode thanks to the electrolytes. Since it does not carry electrons in the process, only ionic movement can occur across the two electrodes while particles move via an exterior circuit.

Separator: Usually composed of a microporous polymeric film, such as a blend of polypropylene (PP) and polyethylene (PE). Lithium ions can travel through the barrier, which physically divides the anode and cathode to avoid short circuits. It is essential for battery safety because it guarantees ionic connectivity while preventing electrical conductivity. Furthermore, because certain substances may melt and clog the openings, the separator acts as a safeguard by stopping the flow of ions if the battery heats up.

Current Collectors: Copper sheets are used to create the anode current collectors. Following expulsion, they gather and move particles across the negatively charged anode to the exterior circuit. Superior conductive properties and stability in the battery's environment are features of the copper sheets. Aluminium foil is used as the material for the electrode current collectors. During recharging, it gathers and moves particles from the exterior network to the positive electrode. Since aluminium does not combine with lithium, it is utilized to ensure efficiency and reliability over an extended period.

Conductive Additives: Usually composed of graphene, tiny carbon nanotubes (CNTs), or carbon black. Conductive additives are used with both the anode and cathode to increase the dielectric materials' electrical resistance and facilitate effective electron transport during both charging and discharging.

Binder: Styrene–butadiene rubber (SBR) or polyvinylidene fluoride (PVDF). The binders hold together the active materials—graphite in the negative electrode and lithium compounds in the cathode—and attach them to the current collectors. While cycling, the glue maintains its mechanical integrity and prevents the substances from coming loose or disintegrating.

Battery Case/Packaging: In accordance with the design, it can be constructed from steel, plastic, or aluminium. Common cell types include prismatic, spherical, and tubular cells. The battery's exterior shell encloses and shields its interior components from external elements, including moisture, unexpected physical events, and mechanical breakdowns. Wrapping in pouch cells may be made more flexible to save weight and space.

Safety Mechanisms: Positive Temperature Coefficient (PTC) toggles, heat fuses, and pressure relief mechanisms are examples of built-in protections. These devices guard against short circuits, excessive charging, and overheating of the battery. Whenever the internal pressure rises too high, pressure-relieving valves evacuate gas, and PTC valves disconnect the system when temperatures reach dangerous levels.

Working Principle: Lithium ions are forced to travel throughout the electrolyte solution from the positive to the negative electrode while charging due to an external electrical current. Through an external connection, electrons can also move from the positive to the negative electrode. The structure of the anode, which is usually made of graphite, stores the lithium ions. Lithium ions release energy by travelling through the electrolytes from the negative electrode to the positive electrode. Using a separate circuit, electrons simultaneously flow from the negative electrode to the cathodes, powering the attached device [18].

The different types of lithium-ion batteries are given below:

i. **Lithium cobalt oxide ($LiCoO_2$)**
ii. **Lithium iron phosphate ($LiFePO_4$)**
iii. **Lithium nickel manganese cobalt oxide (NMC)**
iv. **Lithium nickel cobalt aluminium oxide (NCA)**
v. **Lithium manganese oxide ($LiMn_2O_4$)**

12.5.1 Chemical reactions

- Anode reaction (during charging):

$$\mathrm{Li} \rightarrow \mathrm{Li}^{+} + \mathrm{e}^{-}$$

- Cathode reaction (during charging):

$$\mathrm{LiCoO_2} \rightarrow \mathrm{Li}(1-x)\mathrm{CoO_2} + x\mathrm{Li}^{+} + x\mathrm{e}^{-}$$

- Anode reaction (during discharging):

$$Li^{+} + e^{-} \rightarrow Li$$

- Cathode reaction (discharging):

$$Li\,(1-x)CoO_2 + xLi^{+} = xe^{-} \rightarrow LiCoO_2$$

- Overall reaction:

$$LiCoO_2 \rightarrow Li(1-x)CoO_2 + xLi$$

Lithium-ion batteries have several benefits, such as high power density, extended lifespan, minimal discharge rate, excellent efficiency, and small size. The primary drawbacks and issues of Li-ion batteries include their high cost, short lifespan, safety risks, and environmental impact, among other actors. Significant advances in battery technology have been made in recent years, driven by the increased demand for electricity storage in clean energy systems, electric vehicles (EVs), and portable devices. Recent developments and active research topics will be discussed. These batteries are used in power tools, healthcare devices, grid storage, electric vehicles, and electronic goods, among other applications [19].

12.6 LITHIUM POLYMER (LiPo) BATTERIES

Such are the kinds of rechargeable battery packs that are becoming more and more common in a variety of applications, such as electric vehicles, remote-controlled devices, and household appliances. Although they are a type of lithium-ion battery, they differ in terms of the packaging of their electrolytes. LiPo batteries, in contrast to faceted or rectangular cells, have a malleable polymer shell that enables them to be produced in a range of sizes and shapes, allowing for use in small and personalized designs. Compared to the liquid electrolytes used in conventional lithium-ion battery packs, this employs a gel-like electrolyte that is more durable. However, for electricity storage and transport, they essentially rely on a similar Li-ion mechanism [20]. In addition, they have a high power density, which allows them to hold more power in a lightweight and smaller packaging than other types of batteries, such as nickel–metal hydride (NiMH) or nickel–cadmium (Ni-Cd). LiPo battery packs are smaller than lithium-ion equivalents because of their flexible construction and absence of a metal shell, which makes them perfect for scenarios where weight is an important consideration. Additionally, they offer extremely high discharge rates, which is crucial for applications like remote control automobiles and drones that require a lot of power in brief bursts. Figure 12.3 depicts

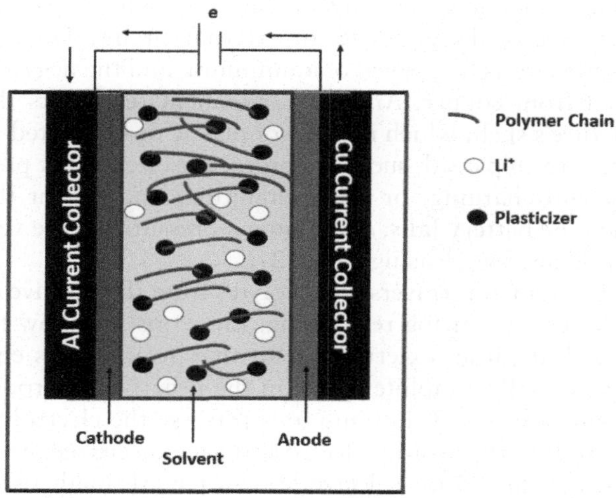

Figure 12.3 Structure of lithium polymer battery.

the structural layout of the lithium polymer battery. The main elements of batteries made from lithium polymer are given below:

i. Cathode (Positive Electrode): Usually made from a lithium–metal oxide, including lithium iron phosphate ($LiFePO_4$), lithium manganese oxide ($LiMn_2O_4$), or lithium cobalt oxide ($LiCoO_2$). An electrical current is produced during discharging when the lithium ion moves across the electrolyte from the negative electrode to the positive electrode. The movement is reversed during recharging, and lithium ions return to the negative electrode.
ii. Anode (Negative Electrode): Composed mostly of carbon-based compounds, such as graphite. During recharging, lithium ions are stored here. Lithium ions move from the negative electrode to the positive electrode when the power source drains.
iii. Electrolyte: A type of polymer that resembles gel and contains lithium salts, such as lithium hexafluorophosphate ($LiPF_6$). It facilitates the transfer of lithium ions across the anode and the cathode during both charging and discharging. LiPo battery packs are more secure and versatile since the electrolytes are contained in solid polymers as opposed to a liquid.
iv. Separator: An extremely thin permeable polymer membrane that separates the negative electrode from the positive electrode to avoid direct physical contact, which might cause an electrical fault. It is frequently composed of polyethylene or polypropylene. Lithium ions can flow through when electrons are blocked.

The remaining components are current collectors, which are used to gather and transport electrical charges to an external circuit. The positive electrode of the collector is composed of aluminium, and the negative electrode is constructed from copper. Another component represents the battery's interior structure's shell, which is made of plastic or laminated aluminium.

Other parts are employed, such as a protective circuit that prevents overcharging, over-discharging, or short-circuiting, and a vent that releases pressure when the battery fails, containing a pressure release vent designed to prevent accidents by releasing gases [21].

Working: Lithium ions traverse electrolytes from the positive electrode to the negative electrode during recharging. Electrons also flow to the negative electrode through an external circuit that accumulates energy in the power source. Usually, graphite sheets in the negative electrode are where the lithium ions are stored. Lithium ions traverse the electrolyte from the negative electrode to the positive electrode during discharge. As the charged particles return to the positive electrode, they pass through an external circuit and supply electricity to the device. When the battery runs out of stored electricity, the procedure is repeated [22].

These batteries have a number of benefits, including high discharge capability, lightweight design, and configurable size and form. The main drawbacks of these battery packs are their shorter life expectancy, susceptibility to degradation, and safety issues. Drones, remote-controlled vehicles, handheld electronics, and electric vehicles are the primary applications for battery packs [23].

12.7 SOLID-STATE BATTERIES

Unlike traditional lithium-ion or lithium polymer battery packs, which employ liquid or gel electrolytes, solid-state battery packs are a more sophisticated kind of reusable battery. Solid-state battery packs have a longer lifespan, better safety, and an increased energy density, among other possible benefits. They are being intensively explored for use in battery storage, electronic goods, and electric vehicles (EVs) [24]. Figure 12.4 illustrates the structural layout of solid-state batteries. The primary components of solid-state batteries are given below:

i. Cathode (Positive Electrode): Lithium-based materials such as lithium cobalt oxide ($LiCoO_2$), lithium iron phosphate ($LiFePO_4$), or lithium nickel manganese cobalt oxide (NMC), which transport lithium ions during the discharge process, are commonly used in solid-state battery packs, much like in lithium-ion batteries. Lithium ions move from the positive electrode towards the negative electrode when the energy source charges.

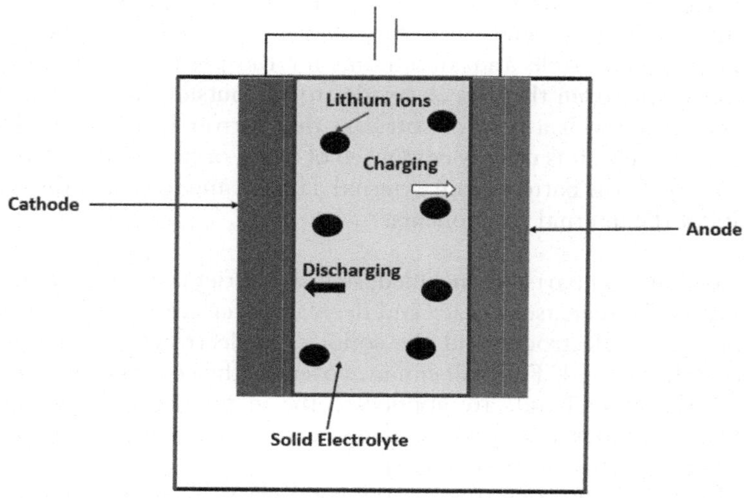

Figure 12.4 Structure of solid-state batteries.

ii. Anode (Negative Electrode): Graphite is typically used as the anode in conventional lithium-ion battery packs. The element lithium is one of the many materials that may be used to create the negative electrode in solid-state battery packs, significantly boosting the battery's power density compared to traditional anode materials. During recharging, they serve to store the lithium ions. Ions pass through the solid electrolytes from the negative electrode to the positive electrode during the battery discharge process, producing electricity.
iii. Solid Electrolyte: A variety of materials, including solid glasses, polymers, and porcelain, can be used to create solid electrolytes. Materials that are often studied include earthenware containing phosphate, oxide, and sulphur. The solid electrolytes facilitate the movement of lithium ions across the two electrodes during the charging and discharge processes, taking the place of the liquid or gel electrolytes in conventional batteries. In addition, it acts as a separator, permitting ion movement while blocking electrical connection between the two electrodes.
iv. Separator: The solid electrolytes themselves serve as a barrier in many situations. If an additional isolation element is required, it is constructed from an impermeable insulating material that allows ions to flow through the solid electrolytes while blocking direct electrical contact between the two electrodes.

v. Current Collectors and Packaging: The purpose of electricity collectors, which are composed of metallic foils (such as copper for the negative electrode and aluminium for the positive electrode), is to move ions from the two electrodes to the outside circuit. The wrapping material is a hard or soft case that surrounds the entire battery architecture. It is usually composed of metal or plastic and serves both to protect the battery from external damage and to ensure the reliability of the internal components.

Certain additional parts are included, such as barrier coatings and intermediate layers. To increase contact and decrease interface resistance between the conducting electrodes and the solidifying electrolyte, interconnected layers can be utilized. This will enhance overall efficiency. The barrier coatings, on the other hand, are applied to the metal electrodes to increase chemical resistance and prevent deterioration through cycling, thereby extending the battery's lifespan [25].

Working: Lithium ions traverse the solid electrolytes from the positive electrode to the negative electrode during recharging. After passing through an external circuit, ions arrive at the surface of the anode and combine with lithium ions there. In the case of batteries, this process stores electrical power. If a lithium metallic anode is used, lithium ions are immediately formed as an extremely thin film of lithium metal. During discharging, lithium ions travel across the solid electrolytes from the negative electrode to the positive electrode. Simultaneously, electrons travel through an external circuit from the negative electrode to the cathode, powering the connected device [26].

Solid-state battery packs have a greater power density, enhanced safety, extended longevity, and quicker charging. Thermal sensitivity, interface resistance, and fabrication difficulty are the main obstacles. These battery packs are used in consumer electronics, aerospace, defence, and electric vehicles (EVs) [25].

12.8 LITHIUM–SULPHUR (Li-S) BATTERIES

In contrast to traditional lithium-ion battery packs, Li-S power sources, a new type of rechargeable battery, have a significantly higher power density. They employ sulphur, which is both abundant and environmentally benign, as the reactive material in the electrode. However, Li-S battery packs still face technical issues, such as capacity decline and shorter shelf life [27]. Figure 12.5 shows the structure of lithium–sulphur battery packs. The main components of Li-S batteries are listed here:

i. Cathode (Positive Electrode): In Li-S batteries, sulphur (S) or compounds based on sulphur are the main materials used as the positive electrode. Because of its low cost, high potential energy content,

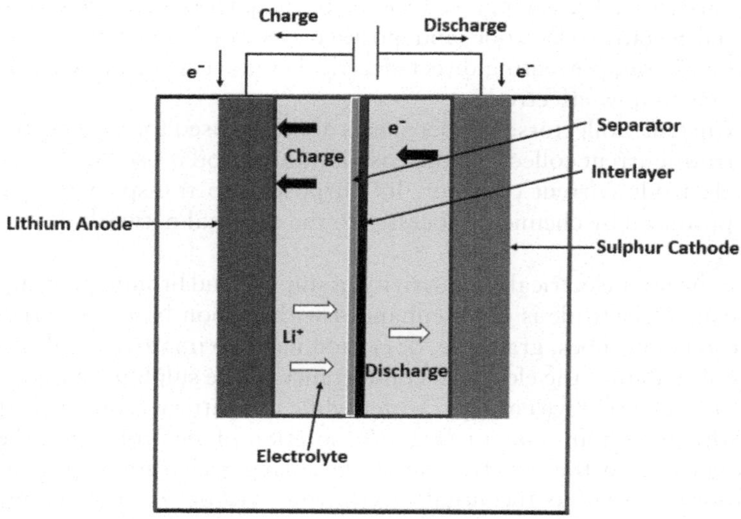

Figure 12.5 Structure of lithium–sulphur battery.

and natural availability, sulphur is a desirable material. Its purpose is to produce energy during discharge; i.e., when sulphur and lithium combine to generate lithium polysulphides (Li_2S_4 to Li_2S), energy is released during the reaction. The process reverses during charging, storing lithium ions in the negative electrode and regenerating sulphur in the positive electrode.

ii. Anode (Negative Electrode): Lithium material is commonly used for the negative electrode. The use of lithium metal, which has a much higher capacity compared to graphite, the negative electrode material in lithium-ion battery packs, contributes to the high power density of Li-S battery packs. Lithium ions are stored in the negative electrode of Li-S batteries during recharging. During discharge, lithium ions move from the negative electrode to the positive electrode, where they combine with sulphur to generate electrical power.

iii. Electrolyte: Generally, a liquid mixture of lithium ions combined with chemical solvents serves as an electrolyte in a Li-S battery. Throughout the charge and discharge cycles, it facilitates the movement of lithium particles through the negative and positive electrodes. Additionally, it helps dissolve lithium polysulphides that are created during cycling; however, this may result in the "polysulphide shuttle effect," in which these intermediate compounds return to the negative electrode and shorten the lifespan and effectiveness of the batteries.

iv. Separator: Typically, a perforated polymer barrier, such as polypropylene (PP) or polyethylene (PE), similar to those used in lithium-ion battery packs, is employed to create the partition. Its role is to keep the negative electrode and positive electrode apart during battery operation, preventing direct electrical contact while allowing lithium ions to flow effectively.

v. Current Collectors: Copper sheets are often used for the negative electrode current collector, whereas aluminium foil is used for the positive electrode current collector. Its purpose is to transport the particles produced by chemical processes to the external network.

Due to the poor electrical conductivity of sulphur and lithium polysulphides, the positive electrode is often enhanced with carbon-based materials such as carbon nanotubes, graphene, or carbon black to improve ion flow. These materials enhance the electrical conductance of the sulphur-based cathode, guaranteeing effective energy transfer while the battery is operating [28].

Working: Lithium ions undergo the solution of electrolyte via the positive electrode to the negative electrode during recharging. Concurrently, electrons are sent to the negative electrode via an external connection. Lithium material is accumulated on the negative electrode as a consequence. However, lithium ions travel through the electrolyte solution from the negative electrode to the positive electrode while discharging. Lithium ions and sulphur in the positive electrode combine to generate lithium sulphide (Li_2S) or lithium polysulphides (Li_2Sn). Devices are powered by the energy released by these processes. The discharging process is completed when electrons flow through an external circuit from the negative electrode to the positive electrode [29].

Li-S batteries provide several benefits, including a higher charge density, affordable materials, and environmental friendliness. However, Li-S batteries face drawbacks and challenges, such as the polysulphur shuttle effect, short cycle life, volume growth, and poor electrical conductivity. Unmanned aerial vehicles, aviation, grid storage, and electric vehicles (EVs) are the main applications for these batteries [30].

12.9 SODIUM-ION BATTERIES

A developing form of renewable battery pack called sodium-ion (Na-ion) power sources is being researched as a substitute for lithium-ion batteries. Na-ion battery packs are a desirable alternative for large-scale power storage and applications where cost and availability are more essential than high energy density, since sodium is cheap, plentiful, and readily accessible. Although they employ sodium ions rather than lithium ions, they have different components and structure and operate on a similar basis to battery packs [31]. Figure 12.6 illustrates how sodium-ion batteries are constructed. The principal parts of sodium-ion battery packs are as follows:

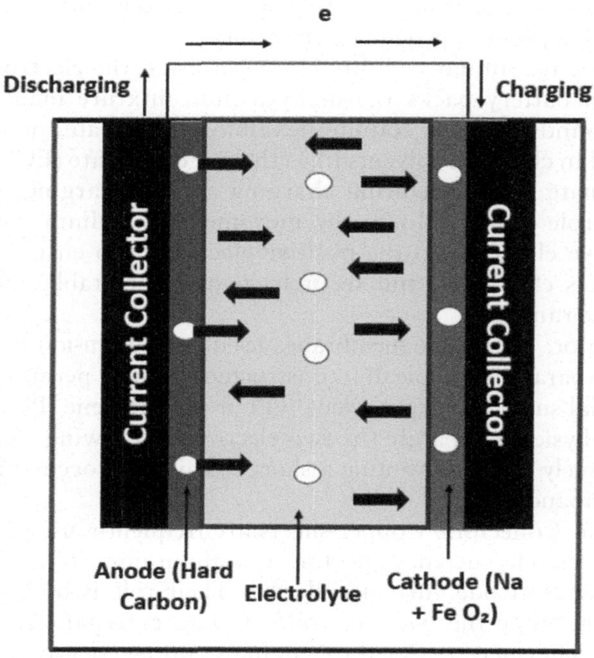

Figure 12.6 Structure of a sodium-ion battery.

i. Cathode (Positive Electrode): Sodium transition metal oxides or polyanionic compounds are commonly utilized as the positive electrode materials in Na-ion battery packs. Sodium iron phosphate ($NaFePO_4$), sodium manganese oxide ($NaMnO_2$), and sodium cobalt oxide ($NaCoO_2$) are examples of frequently used substances. These battery packs work by moving sodium ions from the positive electrode to the negative electrode via the electrolyte solution while electrons travel through an external circuit to supply electricity to the connected device during discharging. The cathode functions as the main source of sodium oxide and plays a significant role in determining the battery's power and energy density.
ii. Anode (Negative Electrode): Sodium-ion battery packs frequently employ materials such as tough carbon or titanium-based substances, including sodium titanate, as negative electrode components, whereas lithium-ion battery packs typically utilize graphite. Although sodium ions are larger than lithium ions and have a harder time intercalating within the graphite lattice, graphite does not work effectively alongside sodium ions. During recharging, it serves as a repository for sodium atoms at the negative electrode. These ions transfer from

the negative electrode to the positive electrode during the battery's discharge process, producing electricity.

iii. Electrolyte: Similar to lithium-ion batteries, the electrolyte used in Na-ion battery packs is usually a fluid mixture made of sodium compounds (such as sodium hexafluorophosphate, or NaPF6) dispersed in chemical solvents like ethylene carbonate (EC) or dimethyl carbonate (DMC). During charging and discharging, the electrolyte's role is to facilitate the movement of sodium ions from the negative electrode to the positive electrode. To ensure the battery operates effectively and securely, it must be stable across a wide voltage range.

iv. Separator: Unlike the membranes used by lithium-ion battery packs, this separator is typically constructed from a permeable polymer material such as polypropylene (PP) or polyethylene (PE). Its purpose is to physically separate the two electrodes, allowing sodium ions to flow freely while preventing a short circuit from occurring in battery performance.

v. Current Collectors: Copper sheets are frequently used for the negative electrode current collector, and aluminium foil is used for the positive electrode current collector. Their role is to transport electrons between the two electrodes and an external circuit and back again. They play a critical role in maintaining effective charging and discharging cycles.

Conductive additives are the additional ingredients. To enhance electrical conductivity and the movement of ions during cell operation, conductive materials such as carbon black or graphite are used in the membranes. This improves the overall battery lifespan. Furthermore, the glue, which is composed of polyvinylidene fluoride (PVDF) or another polymer-based material, serves to maintain the mechanical strength of the conducting electrodes, ensuring they retain their dimensional stability during recharging. It holds the active materials and conductive components within the electrode's structure [32].

Working Principle: Na ions traverse the electrolyte solution from the negative electrode to the positive electrode during discharging, liberating the stored electricity that drives an external device. The electric power is produced by electrons moving through the external circuitry. Na particles, on the contrary, are pulled out of the positive electrode and placed into the negative electrode, where they will be preserved during recharging. During the recharging process, ions as well as electrons travel in the opposite direction, replenishing the battery's stored energy [33].

The benefits of Na-ion batteries include the abundance of sodium, lower cost, environmental friendliness, and thermal stability. The challenges and disadvantages include lower energy density, larger ion size, shorter cycle life, and limited commercial maturity. Na-ion batteries are used in consumer electronics, portable electricity storage, electric vehicles (EVs), electrical grid storage, and other applications [33].

12.10 LEAD ACID BATTERIES

Among the first types of rechargeable batteries, lead acid battery packs are frequently used for independent electricity storage, power backups, and vehicles. Because of their reliability, cost-effectiveness, and ability to provide large surge currents, these batteries continue to be widely used, even though their power density is lower than that of contemporary battery technologies such as lithium-ion [34]. Figure 12.7 depicts the construction of a lead acid battery. The principal parts of a lead acid battery are as follows:

i. Cathode (Positive Electrode): Lead dioxide (PbO_2) is placed onto a lead plate to form the positive electrode. When lead sulphate ($PbSO_4$) is formed during discharging, PbO_2 reacts with sulphuric acid, delivering electricity to the external circuit. Lead dioxide is produced during the charging cycle when the reaction is reversed.
ii. Anode (Negative Electrode): Porous lead (Pb), which is also placed onto a lead plate, serves as the negative electrode. It works during discharging by liberating ions when lead sulphate ($PbSO_4$) is formed by the reaction of porous lead and sulphuric acid. Lead sulphate is transformed back into spongy lead during recharging.

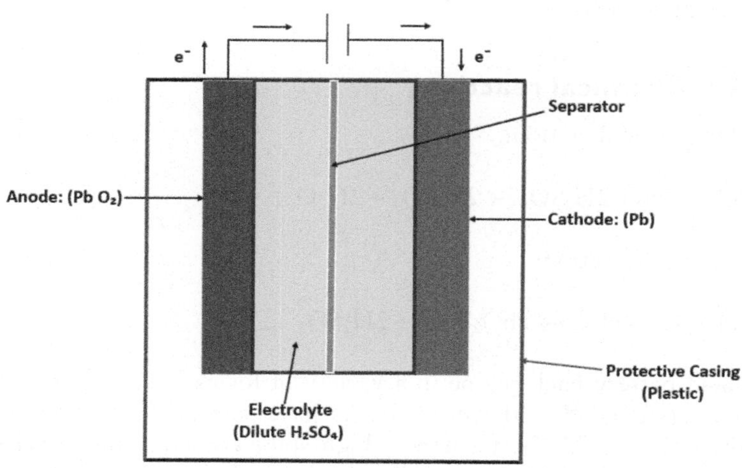

Figure 12.7 Structure of a lead acid battery.

iii. Electrolyte: Sulphuric acid (H_2SO_4) dissolved in water acts as the electrolyte. Throughout the discharging and recharging cycles, H_2SO_4 serves as the medium for the flow of electrons between the positive and negative electrodes. It actively participates in the energy-storing and -releasing chemical processes.
iv. Separator: Glass fibre, polyethylene, or microporous polymers are frequently used to make separators. The separator's role is to keep the two pairs of electrodes apart so that electrons in the solution can flow freely and to prevent short circuits in the battery.
v. Battery Case: Usually, polypropylene or another acid-resistant polymer is used for the outer shell. It holds the solution and serves to safeguard the interior parts. It is designed to be strong and impervious to the destructive properties of sulphuric acid.
vi. Current Collectors: The electric current receivers in both the positive and negative plates are grids made of lead. Their role is to allow ions to flow back to the surfaces during recharging and to transfer electrical energy through the surfaces to the external circuit during discharging.

Working Principle: PbO_2 and H_2SO_4 combine at the positively charged electrode to generate $PbSO_4$ and release energy to the external circuit. It releases electrons when it combines with H_2SO_4 at the opposite end of the electrode (Pb) to generate $PbSO_4$. The device connected to the power source is powered by electrons moving through the external circuit. The reverse process occurs during recharging: H_2SO_4 is restored in the electrolyte solution and $PbSO_4$ is generated on each plate. This is transformed back into lead (Pb) at the positive plate and PbO_2 at the negative plate [35].

12.10.1 Chemical reactions

- Discharge Reaction:

$$PbO_2 + Pb + 2H_2SO_4 \rightarrow 2PbSO_4 + 2H_2O$$

- Charge Reaction:

$$2PbSO_4 + 2H_2O \rightarrow PbO_2 + Pb + 2H_2SO_4$$

Lead acid battery packs come in a variety of forms, including sealed and flooded types. Lead acid battery packs have several advantages, including low cost, reliable performance, high spike current, and recyclability. However, they also have several drawbacks, including low power density, limited lifespan, high weight, and the need for ongoing maintenance. Lead acid battery packs are used in isolated energy storage, uninterruptible power supplies (UPS), and electric vehicles (EVs) [36].

12.11 NICKEL–CADMIUM (Ni-Cd) BATTERIES

Rechargeable battery packs of the nickel–cadmium (Ni-Cd) variety have been around for a long time and are employed in various industries, such as power equipment, aircraft, and mobile devices. Although technological advances like lithium-ion (Li-ion) and nickel–metal hydride (NiMH) battery packs have largely supplanted Ni–Cd rechargeable batteries for numerous applications, Ni–Cd battery packs remain valued for their longevity, reliability, and ability to perform well in extreme temperatures [37]. Figure 12.8 depicts the construction of a nickel–cadmium battery. The principal components of Ni–Cd batteries are given below:

i. Cathode (Positive Electrode): The material used on the positively charged plate is nickel oxyhydroxide (NiO_2H). NiO_2H serves to react with ionized hydroxide in the electrolyte solution to generate nickel hydroxide ($Ni\ (OH)_2$) when discharging. During charging, this reaction is reversed, converting nickel hydroxide back to nickel oxyhydroxide.
ii. Anode (Negative Electrode): Cadmium (Cd) is used to make the negatively charged plate. Cadmium hydroxide ($Cd\ (OH)_2$) is created during discharging when cadmium combines with hydroxide particles,

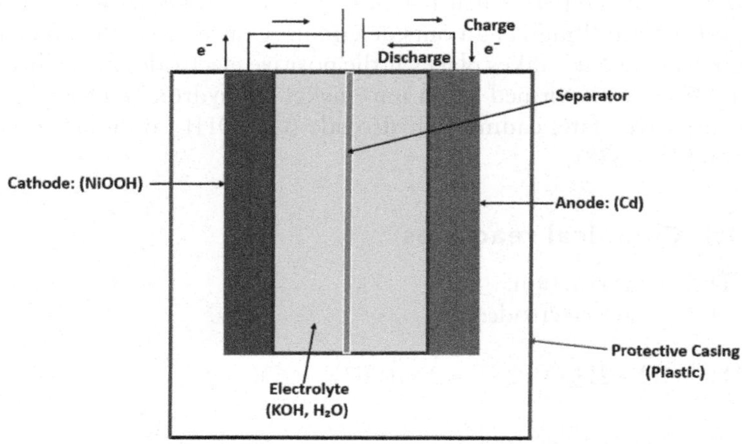

Figure 12.8 Structure of a nickel–cadmium battery.

delivering ions into the external circuit. Cadmium hydroxide is converted back into cadmium (Cd) after recharging.

iii. Electrolyte: In Ni–Cd battery packs, potassium hydroxide (KOH), an effective alkaline solution, is commonly used as the electrolyte. During both processes, the electrolyte's role is to facilitate the transfer of hydroxide ions (OH^-) across the positive and negative plates. While it aids in the transport of ions required for the power source to function, the electrolyte solution generally does not participate in oxidation processes.

iv. Separator: Typically, polyamide or a microporous plastic is used to create the separator. This separator physically divides both electrodes to avoid short circuits, allowing ions in the salt solution to flow freely.

v. Battery Case: Usually composed of sturdy plastic or metallic material, such as nickel-plated steel, the exterior casing serves to protect the plates, electrolytic solution, and other internal components, safeguarding the battery from damage and minimizing leakage of the electrolytes.

vi. Current Collectors: In Ni–Cd battery packs, current collectors made of nickel-plated metal or other conductive materials are employed. Their purpose is to conduct electricity through a battery's terminals and into an external circuit.

Working Principle: Nickel hydroxide (Ni $(OH)_2$) is formed at the positive plate by the reaction of nickel oxyhydroxide (NiO_2H) with H_2O and hydroxide particles (OH^-) that constitute the solution. This reaction releases energy into the outside circuit. Cadmium (Cd) combines with hydroxide particles (OH^-) in the solution at the negative plate to generate cadmium hydroxide (Cd $(OH)_2$), which releases ions. The linked device is powered by electrons travelling via an outside circuit. Conversely, when recharging, the opposite reaction takes place: at the positive electrodes nickel hydroxide (Ni $(OH)_2$) is transformed again into nickel oxyhydroxide (NiO_2H), while at the negative plate, cadmium hydroxide (Cd $(OH)_2$) diminishes again to cadmium (Cd) [38].

12.11.1 Chemical reactions

- Discharge reaction:
  - Positive electrode:

$$2NiOOH + 2H_2O + 2e^- \rightarrow 2Ni(OH)_2 + 2OH^-$$

  - Negative electrode:

$$Cd + 2OH^- \rightarrow Cd(OH)_2 + 2e^-$$

- Charge reaction:
  - Positive electrode:

$$2Ni(OH)_2 + 2OH^- \rightarrow 2NIOOH + 2H_2O + 2e^-$$

  - Negative electrode:

$$Cd(OH)_2 + 2e^- \rightarrow Cd + 2OH^-$$

Ni–Cd battery packs have a long lifespan, are immune to excessive charging, have a high drain rate, and can withstand a broad temperature variation. The drawbacks include discharge rate, toxic effects, memory effect, decreased power density, expense, and environmental issues, among other things. Ni–Cd battery packs are used for electrical tools, aircraft, motor vehicles, emergency illumination, handheld devices, spare power supplies, and other applications [39].

12.12 NICKEL METAL HYDRIDE (NiMH) BATTERIES

In comparison to Ni–Cd battery packs, nickel–metal hydride (NiMH) battery packs are renewable and frequently found in electric automobiles, power equipment, electronic devices, and other applications where a higher energy density is needed. Compared to Ni–Cd battery packs, NiMH battery packs have a higher power density and can be less hazardous, making them a greener choice. Their overall efficiency is enhanced by the fact that they also exhibit fewer memory effects [40]. Figure 12.9 depicts the

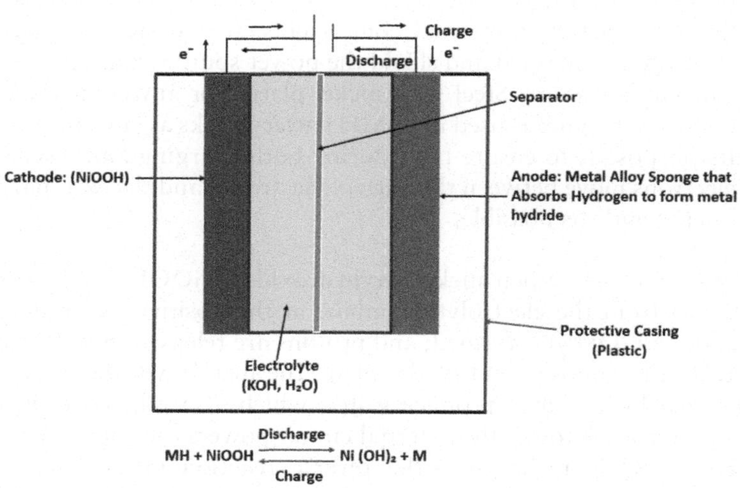

Figure 12.9 Structure of a nickel–metal hydride (NiMH) battery.

composition of nickel–metal hydride battery packs. The principal components of NiMH batteries are given below:

i. Cathode (Positive Electrode): The identical compound utilized in Ni–Cd battery packs, nickel oxyhydroxide (NiOOH), is employed to make the positive plate. The purpose of this discharge process is to generate nickel hydroxide ($Ni(OH)_2$) whenever nickel oxyhydroxide interacts with the solvent, liberating electrons. The method of charging reverses this process.
ii. Anode (Negative Electrode): The metallic hydride material, also known as the hydrogen-storing alloy, that forms the negative plate of NiMH battery packs is usually a combination of nickel, cobalt, and various elements, including lanthanum, cerium, neodymium, and praseodymium. Protons released by the metallic hydride during discharging interact with hydroxide ions in solution. A metallic hydride is formed when an element absorbs hydrogen during charging.
iii. Electrolyte: Potassium hydroxide (KOH) in an aqueous solution is commonly used as an electrolyte. Throughout both the charging and discharging cycles, its role is to facilitate ion movement across the positive and negative electrodes. Moreover, it aids in the passage of hydroxide ions (OH^-) within the cell.
iv. Separator: Microporous polyester, nylon, or another material that allows ions to pass through but prevents direct contact between the two electrodes is used for the separator. Its role is to actively keep the charged particles in the electrolyte from easily moving across both the positive and negative electrodes to avoid short circuits.
v. Battery Case: Usually composed of nickel-plated metal or another robust substance, the outer shell serves as a support structure and holds the battery's internal components. It is designed to keep the electrolyte contained and shield the power source from direct harm.
vi. Current collectors: Steel with nickel plating or any comparable conductive substance is used in NiMH battery packs as current collectors. Its purpose is to ensure that, during both charging and discharging, electrons move between the battery electrodes and the external circuit as efficiently as possible.

Working Principle: When nickel oxyhydroxide (NiOOH) and hydroxide ions (OH^-) from the electrolyte combine at the positive electrode, nickel hydroxide ($Ni\ (OH)_2$) is formed, and protons are released into the external circuit. At the opposite end of the plate, the metal hydride (MH) generates ionized hydrogen to produce water, which also releases energy. The flow of electrons through the external circuit powers the connected device. Conversely, while recharging, the reverse reaction takes place. Nickel hydroxide ($Ni\ (OH)_2$) converts to nickel oxyhydroxide (NiOOH) at the positive plate. A metal hydride (MH) is formed at the negatively charged plate when hydrogen is consumed by the metallic alloy [41].

12.12.1 Chemical reactions

- **Discharge Reaction:**
  - **Positive Electrode:**

$$2NiOOH + 2H_2O + 2e^- \rightarrow 2Ni(OH)_2 + 2OH^-$$

  - **Negative Electrode:**

$$MH + OH^- \rightarrow M + H_2O + e^-$$

- **Charge Reaction:**
  - **Positive Electrode:**

$$2Ni(OH)_2 + 2OH^- \rightarrow 2NiOOH + 2H_2O + 2e^-$$

  - **Negative Electrode:**

$$M + H_2O + e^- \rightarrow MH + OH^-$$

Greater power density, environmental friendliness, reduced memory effect, extended lifespan, safety, and reliability are among the advantages of NiMH battery packs. The disadvantages include a high self-discharge rate, lower energy density than lithium-ion, sensitivity to overcharging, heat generation, etc. The applications of NiMH batteries include consumer electronics, hybrid electric vehicles (HEVs), power tools, medical devices, emergency lighting, etc. [42].

12.13 FLOW BATTERIES

Flow battery packs are a type of rechargeable battery that generate power by pumping liquid electrolytes—stored in external tanks—through the cell. In contrast to traditional batteries, which store electricity within the plates, flow battery packs store power in a liquid state, making it easy to scale up by simply increasing the number of electrolyte containers. Flow battery packs are particularly well-suited for large-scale energy storage applications, including battery backup systems, grid stability, and the integration of renewable energy sources [43]. Figure 12.10 depicts the construction of the flow battery. The principal components of flow batteries are as follows:

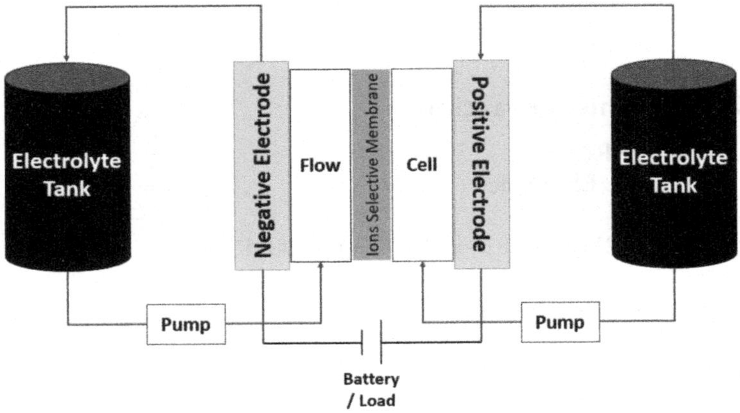

Figure 12.10 Structure of flow batteries.

i. Electrolyte Solutions: Two aqueous solutions of electrolytes that are kept in different tanks are used in flow battery packs. Such solutions typically consist of metallic ions or similar redox-active molecules and are water-based. Examples of such battery packs include zinc bromine flow battery packs, which utilize zinc and bromine, and vanadium redox flow battery packs (VRFBs), which employ vanadium particles in various states of oxidation. The role of the electrolyte in power-storing redox couples is to preserve energy. Energy can be retained or supplied during charge or discharge because the charged particles in the solution become oxidized or reduced at the terminals.
ii. Electrodes: Usually, conductive substances like metallic wire mesh, charcoal felt, or carbon-based paper are used to make these electrodes. They act as sites for redox processes. Reduction processes occur at the positive electrode (cathode), whereas oxidation occurs at the opposite electrode (anode). Through the external circuitry, these processes allow electrons to move and generate electricity.
iii. Ion-selective membrane (Separator): The membrane is usually made of neflon or other ion-conducting polymers. The ion-selective membrane separates the two electrolyte solutions while allowing specific ions (e.g. protons or other ions) to pass through to balance the charge during operation. It prevents the mixing of the two electrolytes while maintaining ionic conductivity, which is essential for the battery's function.
iv. Electrolyte tanks: Tanks are made of materials that resist chemical corrosion, typically plastic, fibreglass, or polymer-lined steel. These tanks

store the electrolyte solutions that hold the active materials. The flow battery's overall energy density is determined by the size of the containers, which allows for expandable storage simply by increasing the tank volume.

v. Pumps: Pumps are typically made from chemically resistant materials such as polypropylene, Teflon, or other corrosion-resistant plastics. The electrolytic mixtures are moved by pumps from storage vessels to the membrane stack, which is the site of the electrochemical processes. The battery's electrical power is influenced by the fluid's circulation rate.
vi. Cell Stack: Each cell is stacked one on top of the other in either parallel or series, with barriers and conductors in between. The cells are housed in frames made of plastic, metal, or other durable materials. The main element in which electrochemical processes occur is the cell's layer. It is made up of several electrochemical components, and the current as well as voltage of the power source are determined by the number of cells present in the stacks.
vii. Current Collectors: These are made from copper, aluminium, or stainless steel, depending on the specific design. Electric power may flow because of the charge collectors, which gather and transfer the electrons generated at the ends of the plates to the external circuit.
viii. Power Electronics: Power electronics ensure the battery's secure and efficient operation by managing the charging and discharging cycles and regulating the supply of electricity to and from the battery.

Working Principle: Using the cell's plates, electrolytic fluids are introduced from the external containers. The energetic species, such as vanadium particles, are oxidized at the negative electrode, liberating electrons. The charged particles travel to the positive electrode, the site of a reduction process, via the external circuit. Electrical energy is produced by the passage of ions across the network and the transport of ions over the protective membrane. In contrast, during the charging cycle, the procedure is inverted, and the battery pack receives electrons from an external power supply. Energy is stored for later utilization via the reduction and oxidation of the electrons in the solution [44].

Scalability, longevity, deep discharge capabilities, low deterioration, ecological friendliness, and other advantages are among the benefits of flow battery packs. Low power density, complicated system layout, reduced efficiency, higher starting costs, and so forth are some of the drawbacks. Grid-scale storage for power, the integration of renewable energy sources, emergency power networks, and remote electricity systems are some of the applications [45].

12.14 ZINC–AIR BATTERIES

Zinc serves as the negative electrode, while airborne oxygen acts as the second electrode in zinc–air battery packs, a subtype of metal–air battery packs. These types of battery packs are remarkable for their low cost, excellent power density, and ecological sustainability, which makes them suitable for a range of uses, such as electricity storage in grids, electric vehicles, and hearing devices. Zinc–air battery packs draw O2 from the atmosphere, which contributes to their excellent energy-to-weight ratio, in contrast to conventional batteries that generally rely on enclosed chemical processes [46]. Figure 12.11 depicts the zinc–air battery's construction. The principal parts of zinc–air batteries are given below:

i. Cathode (Positive Electrode): To maximize the O_2 reduction process (ORR), the positive electrode is usually composed of a permeable carbon substance or a graphite-based composite covered with a metal catalyst (such as compounds of silver, platinum, or cobalt). Upon expulsion, the oxygen that the cathode takes from the atmosphere combines with H_2O and ions to generate hydroxide ions (OH^-). Because the air from the positive electrode allows the battery to "exhale" O_2 from the surrounding air, this process is necessary for the device to operate.
ii. Anode (Negative Electrode): Zinc is used to make the negative electrode; it is usually delivered in the form of a solid or powdered plate. Its role during discharging is to cause zinc to oxidize and release electrons when it combines with hydroxide ions (OH^-) from the electrolytes. This is known as zinc oxide (ZnO). Electricity is produced by this process, powering external equipment.

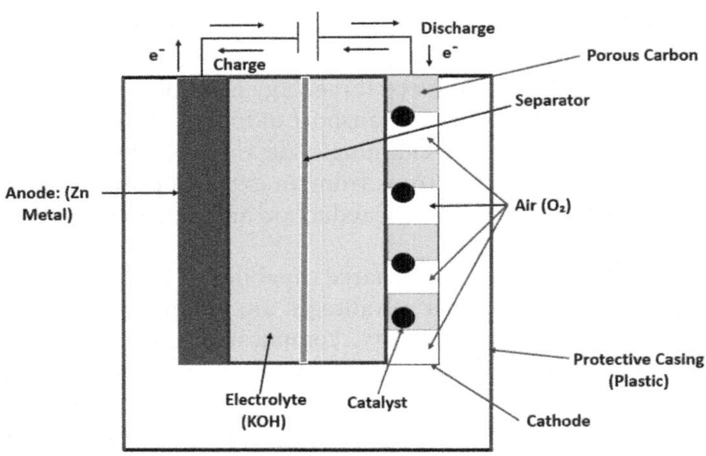

Figure 12.11 Structure of zinc–air batteries.

iii. Electrolyte: Typically, an acidic solution made up of H_2O, such as potassium hydroxide (KOH) or sodium hydroxide (NaOH), serves as the mediator. During the charging and discharging processes, the electrolyte facilitates the movement of hydroxide ions (OH^-) across the negative electrode and positive electrode. It is necessary for the oxidation processes at both plates.
iv. Separator: A polymer substance or a microporous barrier constitutes the element that separates the electrodes. While permitting electrical conductivity, the barrier physically divides the zinc electrode from the positive electrode to avoid shorts in the circuit. Hydroxide ions can flow across it, whereas both electrodes remain apart.
v. Air Electrode (Air Diffusion Layer): This layer is composed of a hydrophobic material (e.g., PTFE, polytetrafluoroethylene) that allows air (oxygen) to enter the battery while preventing water from leaking out. The air diffusion layer enables oxygen from the environment to diffuse into the battery, providing the oxygen needed for the reaction at the cathode. It also helps maintain proper moisture levels inside the battery, preventing it from drying out.
vi. Casing: The casing material is made of plastic or metal, and often includes air vents to allow oxygen to reach the cathode. The casing holds the internal components together and protects them from physical damage. It provides a structure for the air vents that allow oxygen to flow into the battery.
vii. Air vents: The air vents allow ambient air to enter the battery, supplying oxygen to the cathode. These vents are typically covered by a tap before the battery is activated to prevent premature degradation.

Working Principle: At the anode, zinc is oxidized to zinc oxide (ZnO), releasing electrons into the external circuit.

$$Zn + 2OH^- \rightarrow ZnO + H_2O + 2e^-$$

Hydroxide ions are created at the positive electrode by the reaction of airborne O_2 with H_2O and protons.

$$O_2 + 2H_2O + 4e^- \rightarrow 4OH^-$$

At the negative electrode, they participate in the chemical reaction with zinc, generating energy in the process. The hydroxide ions produced at the positive electrode traverse the electrolyte. On the other hand, during recharging, the zinc oxide (ZnO) at the anode is reduced back to zinc, and oxygen is evolved at the cathode [47].

The chief benefits of zinc–air battery packs include greater power density, low cost, environmental friendliness, and stable discharge voltage.

The disadvantages are limited rechargeability, air depletion, water management issues, and lower power output. Applications include hearing aids, electric vehicles, grid energy storage, and backup power [48].

12.15 ALUMINIUM–AIR BATTERIES

Aluminium serves as the negative electrode while airborne oxygen acts as the positive electrode in metal–air battery packs, also known as Al–air battery packs. Their exceptionally high power density and potential for lightweight, inexpensive energy storage are well known. Although work is being done to create rechargeable variants, aluminium–air batteries are not rechargeable (primary) cells. Because of their theoretically higher energy density compared to conventional lithium-ion battery packs, they are especially intriguing for grid-scale power storage and electric vehicles (EVs) [49]. Figure 12.12 depicts the construction of the Al–air battery. The following are the principal parts of Al–air power sources:

i. Cathode (Positive Electrode): The positive electrode is composed of a metallic mesh or permeable carbon coated with an antioxidant (often platinum, as well as silver or other compounds) to aid in the O_2 reduction process. Ions of hydroxide (OH^-) are produced when O_2 from the atmosphere reacts with H_2O and protons from an external circuit through the positive electrode. The battery's operation is powered by this response.

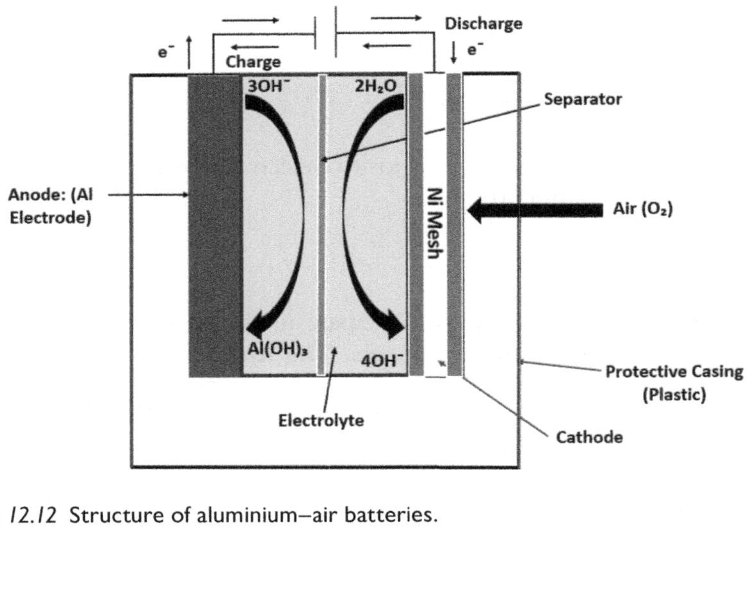

Figure 12.12 Structure of aluminium–air batteries.

$$O_2 + 2H_2O + 4e^- \rightarrow 4OH^-$$

ii. Anode (Negative Electrode): The anode is made of aluminium, typically in the form of a solid plate or sheet. During discharge, aluminium undergoes oxidation, releasing electrons that generate electrical current. The reaction produces aluminium hydroxide as a by-product.

$$4Al + 3O_2 + 6H_2O \rightarrow 4Al\,(OH)_3$$

iii. Electrolyte: To supply hydroxide ions to these processes, a buffer is usually an alkaline liquid such as potassium hydroxide (KOH) or sodium hydroxide (NaOH). The solution facilitates the movement of hydroxide particles (OH^-) between the negative and positive electrodes, enabling the oxidation processes that are required for the batteries to produce electricity.
iv. Separator: A microporous polymer barrier or fibre material forms the partition. Ions can flow freely while physically separating the aluminium anode and air cathode to avoid short circuits. Only hydroxide ions are allowed to pass freely between the two electrodes thanks to the separator.
v. Air Cathode (Air Diffusion Layer): The air diffusion layer is typically made of a hydrophobic material like PTFE (Teflon) or similar materials that allow oxygen to diffuse in while preventing water leakage. This layer enables oxygen from the air to enter the battery, which is essential for the oxygen reduction reaction at the cathode. It also helps maintain proper moisture levels in the electrolyte.
vi. Current Collectors: Made from conductive materials like aluminium, stainless steel, or copper, these are placed on both the anode and cathode sides. The current collectors transfer the electrons generated at the aluminium anode through the external circuit, creating usable electrical power.
vii. Casing: The casing, which is usually composed of metals or plastics, is designed to contain the internal components and shield them from harm. The casing holds the battery components together and provides structure for air vents that allow oxygen to reach the cathode.
viii. Air Vents: Air vents allow ambient air (oxygen) to flow into the battery. These are important for promoting the O_2 elimination process at the positive electrode, as the power source requires a steady input of O_2 from the environment.

Working Principle: When hydroxyl ions (OH^-) from the solution react with aluminium at the negative electrode, aluminium undergoes oxidation to generate aluminium hydroxide ($Al\,(OH)_3$) and releases electrons.

$$Al + 3OH^- \rightarrow Al(OH)_3 + 3e^-$$

Hydroxide particles (OH^-) are created at the positive air electrode when O_2 from the air combines with H_2O and protons from the external circuit.

$$O_2 + 2H_2O + 4e^- \rightarrow 4OH^-$$

The electrons flow through the external circuit, providing electrical power, while the hydroxide ions move through the electrolyte from the cathode to the anode. Aluminium hydroxide ($Al\ (OH)_3$) is formed as a by-product at the anode. This material can accumulate in the battery and may need to be removed or processed depending on the application [50].

The advantages of aluminium batteries include high energy density, low cost, environmental friendliness, and lightweight construction. On the other hand, their disadvantages are non-rechargeability, limited power output, anode corrosion, by-product management, and air and water management. Applications include electric vehicles (EVs), backup power systems, military applications, and grid energy storage [51].

12.16 QUANTUM BATTERIES

Quantum batteries are an emerging concept in energy storage that leverages quantum mechanical principles to store and release energy in ways that are fundamentally different from classical batteries. Quantum battery packs use quantum processes to produce energy, whereas conventional battery packs rely on chemical-based processes, superposition, and entanglement to potentially enable faster charging times, higher energy efficiency, and energy storage capabilities at the nanoscale. Although quantum batteries are still in the theoretical and experimental stages, and there are no commercial quantum batteries available yet, research suggests they could revolutionize energy storage in the future, especially in fields like quantum computing and ultra-fast electronics [52]. The main components of quantum batteries include

i. Quantum Systems: Quantum systems such as quantum dots, ions, or superconducting circuits can serve as the basis for quantum energy storage. These quantum systems store energy in their quantum states. A qubit or quantum dot can exist in multiple energy states simultaneously (thanks to superposition), and entanglement can allow for collective behaviour in energy transfer and storage.

   Quantum Phenomena:

   - Superposition: Allows the quantum battery to exist in multiple energy states simultaneously, increasing energy efficiency.

- Entanglement: Can enable faster charging times by allowing energy to be stored and transferred collectively across multiple quantum systems.

ii. Quantum Charger (Coherent Light or Another Quantum Source): A quantum charger could be a laser or a coherent source of photons, electromagnetic fields, or other quantum systems designed to convert energy into the quantum battery. This component is responsible for charging the quantum battery by exciting the quantum systems (qubits or quantum dots) to higher energy states. The charging process may leverage quantum coherence to accelerate charging speeds beyond what is possible in classical systems.

iii. Quantum Reservoir: The quantum reservoir could be another set of quantum systems or environmental quantum fields that interact with the battery's components. This serves as the environment from which energy is drawn or to which it is released during the recharging and draining process. The reservoir is also designed to optimize the interaction with the battery's quantum systems to minimize energy loss and decoherence.

iv. Energy Extraction Mechanism: In a quantum battery, energy could be extracted via interactions between the quantum system and a classical system (such as an electronic circuit) or through quantum measurement processes. This mechanism enables the retrieval of the stored energy from the quantum states, converting it into usable electrical energy or other forms. It involves manipulating the quantum states (e.g., via controlled de-excitation) to release the stored energy efficiently.

v. Quantum Control Systems: These could be classical or quantum-based control circuits that manipulate the states of the qubits or quantum dots. This component controls the energy storage and release process by precisely managing the quantum states of the system. Quantum control systems ensure that energy is stored and retrieved with minimal loss and maximum efficiency, using algorithms that manage entanglement and coherence.

Working Principle [53]:

- Energy Storage: A quantum battery stores energy by exciting quantum systems, such as qubits or quantum dots, into higher energy levels. Thanks to quantum superposition, these systems can exist in multiple energy states at once, increasing energy capacity. Quantum entanglement between multiple quantum systems can further enhance the battery's storage potential.
- Charging: The battery can be recharged using a coherent energy source (such as a laser or electromagnetic field) that interacts with the quantum systems, pushing them to higher energy states. Quantum entanglement can enable simultaneous charging of all

units, leading to significantly faster charge times compared to classical batteries.

- Discharging: When energy is needed, the quantum states are manipulated to release the stored energy in a controlled manner. The quantum systems de-excite, and the energy is converted into usable electrical energy (or other forms, depending on the application).

The advantages of quantum batteries include ultra-fast charging, high energy efficiency, nanoscale applications, and scalability. The challenges/disadvantages include decoherence, technological immaturity, complexity, and limited applications. The potential application areas of quantum batteries include quantum computing, nanotechnology, fast-charging devices, and medical devices etc [54].

12.17 PAPER-BASED FLEXIBLE BATTERIES

Paper-based flexible batteries are an emerging form of energy storage that uses paper or paper-like materials as a substrate for battery components. These batteries are ideal for use in portable electronics, waste devices, and other areas where typical rigid battery packs are not appropriate as they are pliable, portable, and environmentally friendly. Paper-based batteries can integrate various chemistries, including lithium-ion, zinc-based, and bio-organic systems, depending on the specific application [55]. The structure of paper-based flexible batteries is presented in Figure 12.13. The main components of these batteries include

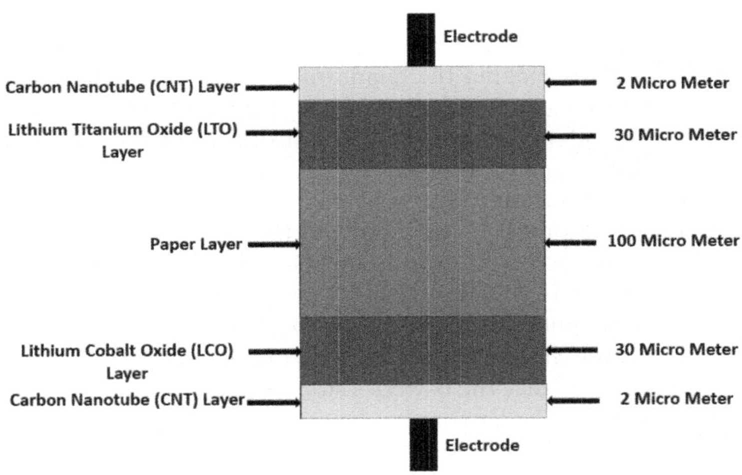

Figure 12.13 Structure of paper-based flexible batteries.

i. Paper Substrate (Flexible Support): The substrate is made of cellulose-based paper or other flexible and biodegradable materials like cotton or polymers. The paper acts as the structural support for the battery's components (electrodes, electrolyte, and separator). It provides flexibility and allows the battery to bend, fold, or roll without losing its integrity. Additionally, the porous nature of paper can allow it to serve as a medium for electrolyte storage.
ii. Cathode (Positive Electrodes): The cathode can be composed of materials like lithium cobalt oxide ($LiCoO_2$), manganese dioxide (MnO_2), or conductive polymers. The positive electrode accepts electrons during the discharging process. Like the anode, the cathode material can be applied as a thin layer onto the paper substrate, allowing for the battery's flexibility.
iii. Anode (Negative Electrodes): The anode can be made of graphite, zinc, or other materials depending on the battery chemistry. It stores electrons during the battery's charging cycle and releases them during discharge. In a paper-based battery, the anode material is often coated or printed directly onto the paper substrate.
iv. Electrolyte: Paper-based batteries can use liquid, solid-state, or gel-based electrolytes, which may include aqueous solutions (like potassium hydroxide), ionic liquids, or polymer-based electrolytes. Throughout the charging and discharging stages, the electrolyte solution facilitates the movement of ions between the two electrodes of the battery. In some designs, the paper itself can be infused with the electrolyte, allowing for the entire battery to be constructed on a single sheet of paper.
v. Separator: The separator can be made from the same paper substrate, or it can be a separate thin porous membrane. Some designs utilize multi-layered paper that acts as both the separator and the electrolyte reservoir. Ions can flow through the partition while it blocks direct contact between the two electrodes, which could result in an electrical short. The paper's natural porous structure can effectively serve this function.
vi. Current Collectors: The current collectors in paper batteries can be made from metallic coatings such as gold, silver, copper, or carbon-based materials like graphene or carbon nanotubes (CNTs). They gather and transfer the electrons produced by the electrochemical reactions in the battery. These collectors are often printed or deposited directly onto the paper in the form of thin, flexible films.
vii. Conductive Links: Conductive inks made from silver nanoparticles, carbon, or graphene can be printed onto the paper. These linkages create the conducting channels that connect the metal plates to the exterior circuitry. Because of the ink's elasticity, the battery can be twisted or deformed without losing functionality.

viii. Encapsulation/Packaging: The battery may be encased in a thin layer of biodegradable plastic, polymers, or laminated films to protect it from external factors such as moisture, oxygen, or mechanical damage. Encapsulation provides protection and ensures the integrity of the battery components while maintaining the flexibility and thinness of the battery.

Working Principle: Throughout the recharging process, protons are transmitted from the positive electrode to the negative electrode via an exterior circuit, while ions flow across the electrolytic solution in the reverse direction, conserving energy in a chemical-based form. The procedure is reversed whenever the power source is in use (discharged). Ions move across the solution to maintain charging balance, whereas protons travel from the negative electrode to the positive electrode via the exterior circuit, activating the attached device [56].

The main advantages of paper-based flexible batteries include their lightweight and thin design, flexibility and foldability, environmental friendliness, cost-effectiveness, versatile applications, and biodegradability. On the other hand, the challenges/disadvantages include lower energy density, scalability, durability, and charge cycles. Applications include wearable electronics, disposable medical devices, RFID tags, smart packaging, bendable displays, and environmental sensors [57].

i. 12.18 Importance of battery technology Powering Modern Electronics: Batteries have enabled the widespread adoption of portable electronics like smartphones, laptops, and wearables. Li-ion technology, in particular, has revolutionized this market by providing lightweight, high-capacity energy storage that powers devices for longer periods [58].
ii. Electric Vehicles (EVs): Because of the urgency to cut GHG pollutants and the rising concerns about global warming, the electrification of public transport has received significant attention lately. At the heart of this shift are batteries, particularly cutting-edge Li-ion and solid-state innovations, which offer the energy storage needed for EVs to compete with conventional gasoline-powered cars in terms of efficiency and capacity.
iii. Renewable Energy Integration: There is an increasing demand for reliable electricity storage as both wind and solar power output rise. With the use of battery systems, excess clean energy produced during peak times can be stored for use when output is low, such as at night and during windless periods. Large storage options are emerging to facilitate the grid's adoption of clean energy sources including sodium-ion battery packs and flow battery packs.

iv. Grid Storage and Backup Power: Battery packs are essential for enhancing energy efficiency, maintaining grid networks, and supplying additional electricity during blackouts. To provide a steady supply of electricity, hospitals, data centres, and other critical infrastructure frequently employ lead acid and lithium-ion technologies.

12.18 CHALLENGES IN BATTERY TECHNOLOGY

Despite their numerous benefits, batteries face several challenges that are the focus of ongoing research [59]:

i. Energy Density Limits: Increasing energy density without sacrificing safety or lifespan is a primary challenge in enhancing the endurance of EVs and the runtime of portable devices.
ii. Safety Concerns: Li-ion batteries, while dominant, pose safety risks such as overheating and fire hazards due to the flammability of liquid electrolytes. Solid-state batteries, which use non-flammable electrolytes, are being developed to address these issues.
iii. Raw Material Sourcing: Substances necessary for the creation of batteries, such as lithium, cobalt, and nickel, are expensive and often sourced from regions with geopolitical or ethical concerns. Developing sustainable and abundant alternatives, such as sodium-ion or lithium–sulphur technologies, is a major focus of research.
iv. Recycling and Sustainability: The rapid growth of battery usage, especially in EVs, has raised concerns about waste management and recycling. Improved recycling processes and the development of batteries made from more environmentally friendly materials are critical for minimizing the environmental impact.

12.19 CONCLUSION

The continued evolution of many industries, including energy preservation and transfer from one place to another, depends heavily on battery power technologies. Battery packs are becoming essential components of electric automobiles (EVs), portable devices, and green energy incorporation as the need for clean energy resources rises. Future developments in battery technology, including increased energy densities, enhanced safety, and extended lifespans, are propelling the transition to a better-energized and ecologically conscious world. Lithium-ion batteries, the currently dominant technology, have enabled significant progress in mobile devices, EVs, and grid storage. However, emerging technologies like solid-state batteries, lithium–sulphur, and sodium-ion offer the potential to overcome existing limitations in energy density, cost, and safety. Additionally, innovations in recycling, sustainability, and alternative materials are addressing environmental concerns associated with raw material sourcing and battery disposal. The continued development of battery energy technology will be critical to achieving global goals related to decarbonization, electrification, and energy efficiency. As research continues, we might anticipate new discoveries that will further reduce costs, enhance performance, and minimize environmental impact, solidifying batteries as a cornerstone of the future energy infrastructure.

REFERENCES

1. Muskan Khan, "Innovations in battery technology: enabling the revolution in electric vehicles and energy storage", *British Journal of Multidisciplinary and Advance Studies*, 5 (1): pp. 23–41, Feb. 2024.
2. George Njema, Russel Ben O. Ouma, Joshua Kibet, "A review on the recent advances in battery development and energy storage technologies", *Journal of Renewable Energy*, 1, pp. 1–35, May 2024.
3. Ahmed Zentani, Ali Almaktoof, Mohamed T. Kahn, "A comprehensive review of developments in electric vehicles fast charging technology", *MDPI, Applied Sciences*, 14 (11), 4728, 2024.
4. Arvind R. Singh, Pradeep Vishnuram, Sureshkumar Alagarsamy, Mohit Bajaj, Vojtech Blazek, Issam Damaj, Rajkumar Singh Rathore, Fahd N. Al-Wesabi, Kamal M. Othman, "Electric vehicle charging technolgies, infrastructure expansion, grid integration strategies, and their role in promoting sustainable e-mobility", *Alexandria Engineering Journal,* Elsevier, Vol. 105, pp. 300–330, Sep. 2024.
5. Franco Maloberti, Anthony C. Davies, "*A Short History of Circuits and Systems*", IEEE, Published by River Publishers, 2016, pp. 1–343, ISBN-978–87–93379-70–1.
6. Julio A. Sanguesa, Vicente Torres-Sanz, Piedad Garrido, Francisco J. Martinez, Johann M. Marquez-Barja, "A review on electric vehicles: technologies and challenges", *MDPI, Smart Cities*, 4 (1), pp. 372–404, Jan. 2021.
7. Santosh Chavan, Bhumarapu Venkateswarlu, Mohammad Salman, Jie Liu, Prakash Pawar, Sang Woo Joo, Gyu Sang Choi, "Thermal management strategies for lithium-ion batteries in electric vehicles- fundamentals, recent advances, thermal models, and cooling techniques", *International Journal of Heat and Mass Transfer,* 232, p. 125918, Sept. 2024.
8. Marina Becker, Ulrich Kunz, Thomas Turek, "Electrically rechargeable zinc-oxygen flow battery with high power density" *Electrochemistry Communication,* 69, pp. 1–17, May 2016.
9. Mckinsey and Company, *Battery 2030: Resilient, Sustainable, and Circular, Article* January 16, 2023. McKinsey & Company. Retrieved from https://www.mckinsey.com/industries/automotive-and-assembly/our-insights/battery-2030-resilient-sustainable-and-circular

10. Anahita Jannesar Niri, Gregory A. Poelzer, Steven E. Zhang, Jan Rosenkranz, Maria Pettersson, Yousef Ghorbani, "Sustainability challenges throughout the electric vehicle battery value chain", *Renewable and Sustainable Energy Reviews*, 191, p. 114176, March 2024.
11. Mehmet Sen, Muciz Ozcan, Yasin Ramazan Eker, "A review on the lithium-ion battery problems used in elecrteic vehicles", *Next Sustainability*, 3, p. 100036, 2024.
12. Brindha Ramasubramanian, JinKiong Ling, Rajan Jose, Seeram Ramakrishna, "Ten major challenges for sustainable lithium-ion batteries", *Cell Reports Physical Science, Cell Press, Science Direct*, 5 (6), p. 102032, June 2024.
13. Gang Zhao, Xiaolin Wang, Michael Negnevitsky, "Connecting battery technologies for electric vehicles from battery materials to management", *iScience*, 25 (2), p. 103744, Feb. 2022.
14. Xin Lai, Yunfeng Huang, Huanghui Gu, Cong Deng, Xuebing Han, Xuning Feng, Yuejiu Zheng, "Turning waste inti wealth: a systematic review on echelon utilization and material recycling of retired lithium-ion batteries", *Energy Storage Materials*, 40, pp. 96–123, Sept. 2021.
15. Marco Piccolino, "The bicentennial of the voltaic battery (1800–2000): the artificial electric organ", *Trends in Neurosciences*, 23 (4), pp. 147–151, April 2000.
16. Ergashali Yuldashevich Rakhimov, Diyorbek Khoshimov, Shuxrat Sultonov, Fozilbek Jamoldinov, "Battery technologies: exploring different types of batteries for energy storage", *BIO Web of Conferences* 84 (10), pp. 3271–3289, Jan. 2024.
17. Takahisa Ohsaki, Norio Takami, Motoya Kanda, Masao Yamamoto, "High performance thin lithium-ion battery using an aluminium-plastic laminated film bag", *Studies in Surface Science and Catalysis*, 132, pp. 925–928, 2001.
18. Neeraj Gupta, Navdeep Kaur, Sanjay K. Jain, Karan Singh Joshal, "Chapter 3 – smart grid power system", In *Advances in Smart Grid Power System: Network, Control and Security*, pp. 47–71, 2021. Elsevier. https://doi.org/10.1016/B978-0-12-824337-4.00009-6
19. F. M. Nizam Uddin Khan, Mohammad G. Rasul, A. S. M. Sayem, Nirmal Mandal, "Maximizing energy density of lithium-ion batteries for electric vehicles: a critical review", *Energy Reports*, 9 (11), pp. 11–21, Oct. 2023.
20. Jayeeta Chattopadhyay, Tara Sankar Pathak, Diogo M. F. Santos, "Applications of polymer electrolytes in lithium-ion batteries: a review", *MDPI, Polymers*, 15 (19), p. 3907, 2023.
21. Kamalhit S. Boparai, Rupinder Singh, "Electrochemical energy storage using batteries, superconductors and hybrid technologies", *Encyclopaedia of Renewable and Sustainable Materials*, 1, pp. 248–254, 2020.
22. Anand Bhatt, Ray Withers FAA, Guoxiu Wang, *"Lithium-ion batteries"*. Australian Academy of Science, March 2016. Retrieved from https://www.science.org.au/curious/technology-future/lithium-ion-batteries
23. Paolo Cicconi, Pradeep Kumar, "Design approaches for li-ion battery packs: a review", *Journal of Energy Storage*, 73 (Part D), p. 109197, Dec. 2023.
24. Abniel Machin, Carmen Morant, Francisco Marquez, "Advancements and challenges in solid-state battery technology: an in-depth review of solid electrolytes and anode innovations", *MDPI, Batteries*, 10 (1), pp. 1–19, 2024.

25. Guolong Lu, Ge Meng, Qian Liu, Ligang Feng, Jun Luo, Xijun Liu, Yang Luo, Paul K. Chu, "advanced strategies for solid electrolyte interface design with MOF materials", *Advanced Powder Materials*, 3 (1), p. 100154, Feb. 2024.
26. Victor A. Agubra, Jeffrey W. Fergus, "The formation and stability of the solid electrolyte interface on the graphite anode", *Journal of Power Sources*, 268, pp. 153–162, 2014.
27. Yelin Deng, Jianyang Li, Tonghui Li, Xianfeng gao, Chris Yuan, "Life cycle assessment of lithium sulphur battery for electric vehicle", *Journal of Power Sources*, 343, pp. 284–295, March 2017.
28. Gabriela Benvenista, Anna Sanchez, Hector Rallo, Cristina Corchero, Beatriz Amante, "Comparative life cycle assessment of li-sulphur and li-ion batteries for electric vehicles", *Resources, Conservation and Recycling Advances*, 15, p. 200086, Nov. 2022.
29. Wai-Yu Ashley Lam, Hong Zhao, Bo Zhang, Li Wang, Hong Xu, Xiangming He, "Significance of direct observation of lithium-ion distribution and potential distribution inside batteries through operando analyses", *Next Energy*, 2, p. 100106, Jan. 2024.
30. Caiwei Wang, Liyun Cao, Jianfeng Huang, Jiayin Li, Koji Kajiyoshi, "Divergent thinking and its application in biomass carbon electrode preparation", *Renewable and Sustainable Energy Reviews*, 138, p. 110564, March 2021.
31. Krishan Arora, Ashok Kumar, Vikram Kumar Kamboj, Deepak Prashar, Bhanu Shrestha, Gyanendra Prasad Joshi, "Impact of renewable energy sources into multi area multi-source load frequency control of interrelated power system", *MDPI Journal of Mathematics*, 9 (2), p. 186, 2021. https://doi.org/10.3390/math9020186
32. Laleh Solhi, Valentina Guccini, Katja Heise, Lina Solala, Elina Niinivaara, Wenyang Xu, Karl Mihhels, Marcel Kroger, Zhuojun Meng, Jakob Wohlert, Han Tao, Emily D. Cranston, Eero Kontturi, "Understanding nanocellulose-water interactions: turning a detriment into an asset", *Chemical Reviews*, 1, pp. 1–13, Feb. 2023.
33. K. M. Abraham, "How comparable are sodium-ion batteries to lithium-ion counterparts?", *ACS Energy Letters*, 1, pp. 1–19, Oct. 2020.
34. P. Kurzweil, J. Garche, "Overview of batteries for future automobiles", In *Lead Acid Batteries for Future Automobiles,* Elsevier, Edited by J. Garche, E. Karden, Patrick.T. Moseley and David A. Rand, pp. 27–96, 2017.
35. Detchko Pavlov, Ahren Knpyeb, M. Stoycheva, Boris Monahov, "Influence of H_2SO_4 concentration on the mechanism of the processes and on the electrochemical activity of the $Pb/PbO_2/PbSO_4$ electrode", *Journal of Power Sources*, 137 (2), pp. 288–308, Oct. 2004. https://doi.org/10.1016/j.jpowsour.2004.06.020
36. Asian Development Bank, "Handbook on battery energy storage system", Manila, Philippines: Asian Development Bank. Dec. 2018. https://www.adb.org/publications/handbook-battery-energy-storage-system
37. Wei Liu, Tobias Placke, K. T. Chau, "Overview of batteries and battery management for electric vehicles", *Energy Reports*, 8, pp. 4058–4084, Nov. 2022.

38. Behnam Zakeri, Sanna Syri, "Electrical energy storage systems: a comparative life cycle cost analysis", *Renewable and Sustainable Energy Reviews*, 42, pp. 569–596, Feb. 2015.
39. Bin Yiu, Xiaorui Liu, Xiayue Fan, Jia Ding, Wenbin Hu, Cheng Zhong, "120 years of nicke-based cathodes for alkaline batteries", *Journal od Alloys and Compounds*, 834, p. 155185, Sept. 2020.
40. Krishan Arora, Gyanendra Prasad Joshi, Mahmoud Ragab, Muhyaddin Rawa, Ahmad H Milyani, Romany F Mansour, Eunmok Yang, "Bilateral contract for load frequency and renewable energy sources using advanced controller", *Computers, Materials & Continua*, 73 (2), 2022, pp. 3165–3180, https://doi.org/10.32604/cmc.2022.026966
41. V. Arun, R. Kannan, S. Ramesh, M. Vijayakumar, "Review on li-ion battery vs nickel metal hydride battery in EV", *Advances in Material Sciences and Engineering*, 1, pp. 1–7, 2022.
42. Pavel M. Rot, Harsh H. Sawant, Pratik P. Shelar, Prashil U. Sarode, S. H. Gawande, "battery health management- a perspective of design, optimization, manufacturing, fault detection, and recycling", *Energy Storage and Saving*, 3 (3), pp. 190–208, Sept. 2024.
43. Suresh Sagadevan, Mohd. Rafie Johan, A. R. Marlinda, Omid Akbarzadeh, Karuppasamy Pandian, M. M. Shahid, Faruq Mohammad, Jiban Podder, "Chapter-1, background of energy storage", In *Advances in Supercapacitor and Supercapattery: An Innovation Toward Energy Storage Devices* (pp. 1–26). Elsevier, 2021. https://doi.org/10.1016/B978-0-12-819897-1.00003-3
44. Thando Juqu, Shane Clayton Willenberg, Keagan Pokaps, Natasha Ross, "Advances in paper-based battery research for biodegradable energy storage", In *Advanced Sensor and Energy Materials*, 1 (4), pp. 1–24, Dec. 2022.
45. Abdul Ghani Olabi, Mohamed Adel Allam, Mohammad Ali Abdekareem, T. D. Deepa, Abdul Hai Alami, Qaisar Abbas, Ammar Alkhalidi, Enas Taha Sayed, "Redox flow batteries: recent development in main components, emerging technologies, diagnostic techniques, large-scale applications, and challenges and barriers", *MDPI, Batteries*, 9 (8), p. 409, 2023.
46. Cian-Tong Lu, Zhi-Yan Zhu, Sheng-Wen Chen, Yu-Ling Chang, Kan-Lin Hsueh, "Effects of cell design parameters on zinc-air battery performance", *MDPI, Batteries*, 8 (8), p. 92, 2022.
47. E. J. Cairns, "Secondary batteries-nickel systems/nickel-zinc", In J. Garche (Ed.), *Encyclopaedia of Electrochemical Power Sources,* Amsterdam, Netherlands: Elsevier, pp. 528–533, 2009.
48. Sudheer Kumar Yadav, Daniel Deckenbach, Jorg J. Schneider, "secondary zin-air batteries: a view on rechargeability aspects", *MDPI, Batteries*, 8 (11), p. 244, 2022.
49. Tai Li, Meng Huang, Xue Bai, Yan-Xiang Wang, "Metal-air batteries: a review on current status and future applications", *Progress in Natural Science: Materials International*, 33 (2), pp. 151–171, 2023.
50. Pradeep Singh, Krishan Arora, Umesh C Rathore, Gyanendra Prasad Joshi, Woong Cho, "Comparative study of controllers in battery energy storage system integrated with doubly fed induction generator-based wind energy conversion system for power quality improvement", *Energy Reports*, 11 (9), 2024, pp. 4587–4600. https://doi.org/10.1016/j.egyr.2024.04.020

51. Giuseppe Antonio Elia, Kostiantyn V. Kravchyk, Maksym, V. Kovalenko, Joaquin Chacon, Alex Holland, Richard G.A. Wills, "An overview and prospective on Al and Al-ion battery technologies", *Journal of Power Sources*, 481, p. 228870, Jan. 2021.
52. Maryam hadipour, Soroush Haseli, "Enhancing the efficiency of open quantum batteries via adjusting the classical driving field", *Results in Physics*, 64, p. 107928, Sept. 2024.
53. Francesco Campaioli, Stefano Gherardini, James Quach, Marco Polini, "Colloquium: quantum batteries", *Review of Modern Physics*, 96 (3), pp. 1–22, July 2024.
54. J. Q. Quach, G. Cerullo, T. Virgili, "Quantum batteries: the future of energy storage?", *Joule*, 7 (10), pp. 2195–2200, Oct. 2023.
55. Amirhossein Ahmadian, Abbas Shafiee, Mohammadali Alidoost, Amin Akbari, "Flexible paper-based li-ion batteries: a review", *World Journal of Engineering and Technology*, 9 (2), pp. 285–299, May 2021.
56. Pradeep Singh, Krishan Arora, Umesh C Rathore, Eunmok Yang, Gyanendra Prasad Joshi, Kwang Chul Son, "Performance evaluation of grid-connected DFIG-based WECS with battery energy storage system under wind alterations using FOPID controller for RSC", *MDPI Journal of Mathematics*, 11 (9), 2023, p. 2100. https://doi.org/10.3390/math11092100
57. Thu H. Nguyen, Arwa Fraiwan, Seokheun Choi, "Paper-based batteries: a review", *Biosensors and Bioelectronics*, 54, pp. 640–649, 2014.
58. Khaled Itani, Alexandre De Bernardinis, "Review on new-generation batteries technologies: trends and future directions", *MDPI, Energies*, 16 (22), p. 7530, 2023.
59. Itani, K., & De Bernardinis, A, Review on newgeneration batteries technologies: Trends and future directions. *Energies*, 16 (22), p. 7530. Basel, Switzerland: MDPI. 2023. https://doi.org/10.3390/en16227530

Chapter 13

Design and control of aircraft model using MATLAB

Suman Lata Tripathi and Deepika Ghai

13.1 INTRODUCTION

Today, aircraft technology as a means of transportation is effectively used for long-distance travel. Its advanced avionics and features allow it to withstand adverse weather conditions, making it a reliable option for addressing transportation challenges in certain parts of the world. At the same time, air travel offers greater speed and convenience than other modes of transportation [1].

Each aircraft type requires distinct characteristics and operational requirements to meet specific criteria, such as size, weight, complexity, intended use, and human factors. These considerations ensure not only the optimal performance of the aircraft but also the safety of the aircraft and its passengers.

The Research Civil Aircraft Model (RCAM) [2] is an actual aircraft model, similar to the Boeing 757-200, designed to simulate a twin-engine civil aircraft. This model comprises six degrees of freedom (6-DoF), which are used to control the aircraft model's surface deflection and throttle position. The three translations pertain to Cartesian coordinates along the x, y, and z axes, while the other three represent the rotations: pitch, roll, and yaw. Additionally, when creating this model, it is necessary to consider parameters such as aircraft mass (kg), aerodynamic chord parameters, and center of gravity (CoG) parameters (Dely, Delx, Delz) [3]. However, this model simplifies certain dynamic variables, such as changing wind and air density, to make the design simple and easier to understand.

In this study, the RCAM is modeled using MATLAB and Simulink, providing various platforms for modeling, simulating, and controlling complex systems. Moreover, utilizing the FlightGear simulator [4] enhances the study by offering an aircraft flying experience in a realistic environment. These tools allow for a comprehensive understanding of aircraft operations.

DOI: 10.1201/9781003542568-13

13.2 TOOL DESCRIPTION

This section explores the applications of an aircraft model, specifically the Research Civil Aircraft Model created using MATLAB and Simulink [5].

13.2.1 MATLAB

MATLAB is a programming language similar to other well-known languages used in the development of scientific and engineering applications, as it operates with its own IDE (Integrated Development Environment) and accessible sets of libraries [6]. The researcher employs MATLAB as the primary software in this work, which includes the design and modeling of the RCAM.

13.2.2 Simulink

Simulink is a graphical programming interface that operates within MATLAB. It is a block diagram environment primarily used to design systems with multidomain models. With this tool, the researcher can more easily understand the complex systems involved in controlling an aircraft through a visual representation of these systems.

13.2.3 FlightGear

It is an open-source flight simulator widely used in research, academic environments, pilot training, and industrial engineering tools. This flight simulator allows the researcher to study and understand the operating principles in the aircraft industry [4].

13.2.4 Methodology/design steps

The RCAM Plant Design is a simplified method for a six degree of freedom (6-DoF) controlled twin-engine aircraft [7]. This design comprises 10 design steps, excluding the initialization of variables. This model sacrifices many dynamic variables (i.e., changing wind and air density) to make the design simple and easy to understand.

13.2.4.1 Input/output vectors

RCAM accepts up to 8 inputs for controls and up to 12 variables for the aircraft states. Tables 13.1 and 13.2 show the *input control parameters and aircraft state variables for the RCAM respectively.* These are listed in the table below:

Table 13.1 Input controls for the research civil aircraft model (RCAM)

Index	*Symbol*	*Name*	*Unit*
$u(1)$	δ_A	Aileron deflection	rad
$u(2)$	δ_E	Elevator deflection	rad
$u(3)$	δ_R	Rudder deflection	rad
$u(4)$	δ_{Th1}	Throttle position of engine 1	–
$u(5)$	δ_{Th2}	Throttle position of engine 2	–
$u(6)$	W_{xE}	Longitudinal wind	m/s
$u(7)$	W_{yE}	Lateral wind	m/s
$u(8)$	W_{zE}	Vertical wind	m/s

Table 13.2 Aircraft state variables for the research civil aircraft model (RCAM)

Index	*Symbol*	*Name*	*Unit*
$x(1)$	u	Velocity in x body-axis	m/s
$x(2)$	v	Velocity in y body-axis	m/s
$x(3)$	w	Velocity in z body-axis	m/s
$x(4)$	p	Roll rate	rad/s
$x(5)$	q	Pitch rate	rad/s
$x(6)$	r	Yaw rate	rad/s
$x(7)$	Φ	Roll angle	rad
$x(8)$	θ	Pitch angle	rad
$x(9)$	ψ	Yaw angle	rad
$x(10)$	x	X position of CoG in F_E	m
$x(11)$	y	Y position of CoG in F_E	m
$x(12)$	z	Z position of CoG in F_E	m

As mentioned, for simplicity's sake, the simulation that will be used will not take into account u(6:8) and thus will be windless. Meanwhile, x(10:12) will undergo a slight change, where instead of the actual position on a flat plane, it will remain as velocity due north, east, and downward, and will be used to calculate the geographical position of the aircraft.

13.2.4.2 Constants

Equations within the model do not just need the variables stated above but also certain constants related to the formulas. Most of these constants depend on the aircraft being modeled, which, in this instance, is closer to a Boeing 757-200. Other constants are natural constants, and some were derived using more complex equations that are not within the scope of this

```
%Nominal vehicle constants
m = 120000;                    %Aircraft total mass (kg)

cbar = 6.6;                    %Mean Aerodynamic Chord (m)
lt = 24.8;                     %Distance by AC of tail and body (m)
S = 260;                       %Wing planform area (m^2)
St = 64;                       %Tail planform area (m^2)

Xcg = 0.23*cbar;               %x position of CoG in Fm (m)
Ycg = 0;                       %y position of CoG in Fm (m)
Zcg = 0.10*cbar;               %z position of CoG in Fm (m)

Xac = 0.12*cbar;               %x position of aerodynamic center in Fm (m)
Yac = 0;                       %y position of aerodynamic center in Fm (m)
Zac = 0;                       %z position of aerodynamic center in Fm (m)

%Engine constants
Xapt1 = 0;                     %x position of engine 1 force in Fm (m)
Yapt1 = -7.94;                 %y position of engine 1 force in Fm (m)
Zapt1 = -1.9;                  %z position of engine 1 force in Fm (m)

Xapt2 = 0;                     %x position of engine 2 force in Fm (m)
Yapt2 = 7.94;                  %y position of engine 2 force in Fm (m)
Zapt2 = -1.9;                  %z position of engine 2 force in Fm (m)

%Other constants
rho = 1.225;                   %Air density @ sea level (kg/m^3)
g = 9.81;                      %Acceleration due to gravity (m/s^2)
depsda = 0.25;                 %Change in downwash wrt alpha
alpha_L0 = -11.5*pi/180;       %Zero lift AoA (rad)
n = 5.5;                       %Slope of linear region of lift slope
a3 = -768.5;                   %Coefficient of alpha^3
a2 = 609.2;                    %Coefficient of alpha^2
a1 = -155.2;                   %Coefficient of alpha^1
a0 = 15.212;                   %Coefficient of alpha^0
alpha_switch = 14.5*pi/180; %alpha boundary of linear and non-linear
```

Figure 13.1 Constants used in the Research Civil Aircraft Model (RCAM).

work. Figure 13.1 shows the constants used in the RCAM. The following are the constants used in the model:

13.2.4.3 Identifying control limits or saturation

In order for the simulation to run optimally, arbitrary limits are placed on the controls of the aircraft. These limits are determined by reference material. They are, however, written in degrees and must be converted to radians. Table 13.3 shows the arbitrary limits applied to aircraft controls.

Table 13.3 Arbitrary limits applied to aircraft controls

Index	*Symbol*	*Name*	*Min*	*Max*
u(1)	δ_A	Aileron deflection	−25°	25°
u(2)	δ_E	Elevator deflection	−25°	10°
u(3)	δ_R	Rudder deflection	−30°	30°
u(4)	δ_{Th1}	Throttle position of engine 1	0°	10°
u(5)	δ_{Th2}	Throttle position of engine 2	0°	10°

Table 13.4 Other computed variables

Name	*Equation*
Air speed	$V_A = \sqrt{u^2 + v^2 + w^2}$
Angle of attack	$\alpha = \tan^{-1}(w/u)$
Angle of sideslip	$\beta = \sin^{-1}(v/V_A)$
Dynamic pressure	$Q = 0.5\ \rho\ V_A^2$

13.2.4.4 Intermediate variables

Since only the variables stated in section (i.) can be used, other necessary variables must be derived from them. Table 13.4 shows the related *computed* variables. There are four important intermediate variables that need to be computed, with their formulas listed below:

13.3 NON-DIMENSIONAL AERODYNAMIC FORCE COEFFICIENTS IN THE STABILITY FRAME

In this design step, the lift, drag, and sideslip force coefficients will be calculated. These values determine the forces that act on the aircraft along the 6-DoF, namely, from the front, back, top, bottom, left, and right of the aircraft. Note that these coefficients act on the stability frame, meaning that only the aerodynamic force impacts the aircraft. The following are the needed equations in this step.

13.3.1 Lift coefficient of the wing and body in the stability frame

$$C_{\mathrm{Lwb}}^{|F_s} = 5.5 * (\alpha - \alpha_{L=0}), \quad \alpha \le 14.5°$$

$$= -768.5\alpha^3 + 609.2\alpha^2 - 155.2\alpha + 15.213, \quad \text{elsewhere}$$

13.3.2 Downwash

$$\varepsilon = \frac{\partial \varepsilon}{\partial \alpha}(\alpha - \alpha_{L=0})$$

13.3.3 Angle of attack of the tail

$$\alpha_t = \alpha - \varepsilon + \delta_E + 1.3 * q * \frac{l_t}{V_A}$$

13.3.4 Lift coefficient of the tail in the stability frame

$$C_{\mathrm{Lt}}^{|F_s} = 3.1 * \frac{S_t}{S} \alpha_t$$

13.3.5 Total lift coefficient in the stability frame

$$C_L^{|F_s} = C_{\mathrm{Lwb}}^{|F_s} + C_{\mathrm{Lt}}^{|F_s}$$

13.3.6 Drag coefficient in the stability frame (neglecting tail)

$$C_D^{|F_s} = 0.13 + 0.07 * (5.5 * \alpha + 0.654)^2$$

13.3.7 Sideslip coefficient in the stability frame

$$C_Y^{|F_s} = -1.6 * \beta + 0.24 * \delta_R$$

13.4 DIMENSIONAL AERODYNAMIC FORCE IN BODY FRAME

To put the calculated aerodynamic force coefficients to use, each will be multiplied by the dynamic pressure and wing platform area and converted to a vector, which will then be multiplied by a rotation matrix to convert to the body frame.

$$\begin{bmatrix} -C_D^{|F_s} \times Q \times S \\ -C_Y^{|F_s} \times Q \times S \\ -C_L^{F_s} \times Q \times S \end{bmatrix}$$

$$F_A^{|F_s} = \begin{bmatrix} \cos(\alpha) & 0 & -\sin(\alpha) \\ 0 & 1 & 0 \\ \sin(\alpha) & 0 & \cos(\alpha) \end{bmatrix}$$

$$C_{S\Rightarrow B}(\alpha) = F_A^{|F_b} = C_{S\Rightarrow B}(\alpha) * F_A^{|F_s}$$

13.4.1 Aerodynamic moment about the aerodynamic center

The following equations will introduce the angular velocities to the aircraft, causing it to roll, pitch, and/or yaw. The angle of attack and angle of sideslip are the primary factors in this step.

$$C_{\text{Mac}}^{|F_b} = \eta + \frac{\partial C_M}{\partial x}\omega_{\text{be}}^{|F_b} + \frac{\partial C_M}{\partial u}[\delta_A;\delta_E;\delta_R]$$

$$\eta = \begin{bmatrix} -1.4 \times \beta \\ -0.59 - \left(3.1 \times \frac{S_t l_t}{S\bar{c}}\right)(\alpha - \varepsilon) \\ 1 - \alpha\beta \end{bmatrix}$$

$$\frac{\partial C_M}{\partial x} = \frac{\bar{c}}{V_A}\begin{bmatrix} -11 & 0 & 5 \\ 0 & -4.03\left(\frac{S_t l_t^2}{S\bar{c}^2}\right) & 0 \\ 1.7 & 0 & -11.5 \end{bmatrix}$$

$$\frac{\partial C_M}{\partial u} = \begin{bmatrix} -0.6 & 0 & 0.22 \\ 0 & -3.1\left(\frac{S_t l_t}{S\bar{c}}\right) & 0 \\ 0 & 0 & -0.63 \end{bmatrix}$$

$$M_{\text{Aac}}^{|F_b} = C_{\text{Mac}}^{|F_b} * Q * S * \bar{c}$$

13.5 AERODYNAMIC MOMENT ABOUT THE CoG

The next step is to convert the equation solved in section (6) to center it around the CoG. To do this, the cross product of the aerodynamic force and the difference of the position vectors is added to the aerodynamic moment about the aerodynamic center.

$$M_{\text{Acg}}^{|F_b} = M_{\text{Aac}}^{|F_b} + F_A^{|F_b} \times \left(\left[0.23\bar{c};0;0.1\bar{c}\right]-\left[0.12\bar{c};0;0\right]\right)$$

13.6 INTRODUCING ENGINE FORCE AND MOMENT

So far, all calculations assume that the aerodynamic force is the only force acting on the aircraft. Now, it's time to introduce the force and moment provided by the engines. It is assumed that the forces from the engines are parallel and equidistant from the x-axis.

$$F_{\text{E1}}^{|F_b} = [\delta_{\text{Th1}} * m * g;\ 0;\ 0]$$

$$F_{\text{E2}}^{|F_b} = [\delta_{\text{Th2}} * m * g;\ 0;\ 0]$$

$$F_{\text{E}}^{|F_b} = F_{\text{E1}}^{|F_b} + F_{\text{E2}}^{|F_b}$$

$$\mu_1 = \begin{bmatrix} X_{cg} - X_{\text{APT1}} \\ Y_{\text{APT1}} - Y_{cg} \\ Z_{cg} - Z_{\text{APT1}} \end{bmatrix}$$

$$\mu_2 = \begin{bmatrix} X_{cg} - X_{\text{APT2}} \\ Y_{\text{APT2}} - Y_{cg} \\ Z_{cg} - Z_{\text{APT2}} \end{bmatrix}$$

$$M_{\text{Ecg1}}^{|F_b} = \mu_1 \times F_{\text{E1}}^{|F_b}$$

$$M_{\text{Ecg2}}^{|F_b} = \mu_2 \times F_{\text{E2}}^{|F_b}$$

$$M_{\text{Ecg}}^{|F_b} = M_{\text{Ecg1}}^{|F_b} + M_{\text{Ecg2}}^{|F_b}$$

13.6.1 Introducing effects of gravity

Of course, the aircraft is still under the influence of Earth's gravity. Thus, a constant downward force is exerted on the aircraft. However, the aircraft's normal is not always coincident with the surface normal, so the gravitational acceleration does not always point downwards at 9.81 m/s².

$$F_G = m * \begin{bmatrix} -g * \sin(\theta) \\ g * \cos(\theta) * \sin(\phi) \\ g * \cos(\theta) * \cos(\phi) \end{bmatrix}$$

13.7 FINALIZING THE STATE DERIVATIVES

After computing the forces and moments that affect the aircraft, it is time to combine them and create a linear equation that will become the derivative of the state variables.

$$F^{|F_b} = F_A^{|F_b} + F_E^{|F_b} + F_G^{|F_b}$$

$$\dot{x}(1:3) = \frac{1}{m} * F^{|F_b} - \left(\omega_{be}^{|F_b} \times V^{|F_b}\right)$$

$$I = m\begin{bmatrix} 40.07 & 0 & -2.0923 \\ 0 & 64 & 0 \\ -2.0923 & 0 & 99.92 \end{bmatrix}$$

$$M_{cg}^{|F_b} = M_{\mathrm{Acg}}^{|F_b} + M_{\mathrm{Ecg}}^{|F_b}$$

$$\dot{x}(4:6) = I^{-1}(M_{cg}^{|F_b} - \left(\omega_{be}^{|F_b} \times \left(I * \omega_{be}^{|F_b}\right)\right)$$

$$H_\phi = \begin{bmatrix} 1 & \sin(\phi)\cos(\theta) & \cos(\phi)\tan(\theta) \\ 0 & \cos(\phi) & -\sin(\phi) \\ 0 & \sin(\phi)/\cos(\theta) & \cos(\phi)/\cos(\theta) \end{bmatrix}$$

$$\dot{x}(7:9) = H_\phi * \omega_{be}^{|F_b}$$

13.8 SIMULINK BLOCK DIAGRAM

Now that the equations for the state derivatives are complete, they will be stacked to form the state derivative vector, which has nine components. The discussed equations are placed inside a MATLAB Function and

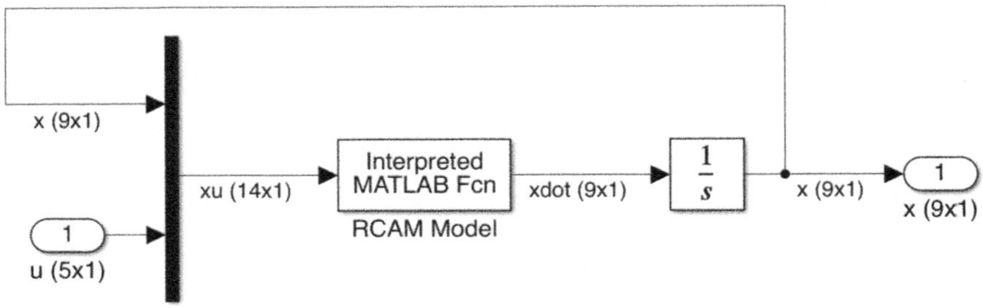

Figure 13.2 State-space simulation with feedback loop.

incorporated into Simulink. The vector is passed through an integrator to obtain the state variables. It forms a feedback loop, serving as both an output and an input of the system. Figure 13.2 shows the *state-space simulation with feedback loop.*

Normally, this would be sufficient to create a comprehensible output where the movement of the state variables is plotted, allowing for a description of the aircraft's overall movement. However, to properly visualize the aircraft's behavior in the model, FlightGear, an open-source flight simulator, will be used. To use FlightGear, the aircraft must first be assigned an actual location using latitude, longitude, and altitude.

13.9 NAVIGATION EQUATIONS

As stated at the beginning of the methodology, the derivative of x(10:12) will be solved, but it will not pass through the integrator. This is because the equations for geodetic position only require the velocity of the aircraft due north, east, and upward. The matrices required to convert the velocities are shown below:

$$C_{1v} = \begin{bmatrix} \cos(\psi) & \sin(\psi) & 0 \\ -\sin(\psi) & \cos(\psi) & 0 \\ 0 & 0 & 1 \end{bmatrix}$$

$$C_{21} = \begin{bmatrix} \cos(\theta) & 0 & -\sin(\theta) \\ 0 & 1 & 0 \\ \sin(\theta) & 0 & \cos(\theta) \end{bmatrix}$$

$$C_{b2} = \begin{bmatrix} 1 & 0 & 0 \\ 0 & \cos(\phi) & -\sin(\phi) \\ 0 & \sin(\phi) & \cos(\phi) \end{bmatrix}$$

$$C_{bv} = C_{b2} * C_{21} * C_{1v}$$

$$C_{vb} = C_{bv}^{'}$$

$$\dot{x}(10:12) = C_{vb} * V^{|F_b}$$

13.10 GEODETIC POSITION

The output x'(10:12) represents velocity due north, velocity due east, and velocity downward, respectively. These will be used in a series of equations that result in the latitude, longitude, and altitude of the aircraft.

13.10.1 Equatorial radius

$$a = 6378137\ m$$

13.10.2 1st eccentricity

$$e = 0.081819190842622$$

13.10.3 Meridian radius of curvature

$$M = \frac{a\left(1-e^2\right)}{(1-e^2\sin(\phi)^2)^{3/2}}$$

13.10.4 Prime vertical radius of curvature

$$N = \frac{a}{(1-e^2\sin(\phi)^2)^{1/2}}$$

13.10.5 Lateral velocity

$$\dot{\phi} = \frac{V_N}{M+h}$$

13.10.6 Longitudinal velocity

$$\dot{\lambda} = \frac{V_E}{(N+h)\cos(\phi)}$$

13.10.7 Altitude climb

$$\dot{h} = -V_D$$

It can be observed that lateral velocity requires an input of altitude, while longitudinal velocity requires an input of longitude. Thus, the geodetic position of the aircraft is also a system with two feedback loops. Therefore, when transferred to Simulink, it will resemble the block diagram shown in Figure 13.3.

The WGS-84 block solves for the radii of curvature, named after the standardization from which the values for the equatorial radius and eccentricity are derived. The function blocks simply carry out the equations stated above, while the latitude and longitude bounds keep their values within -90° to 90° and -180° to 180°, respectively. Figure 13.4 shows a complete block diagram in Simulink of the RCAM.

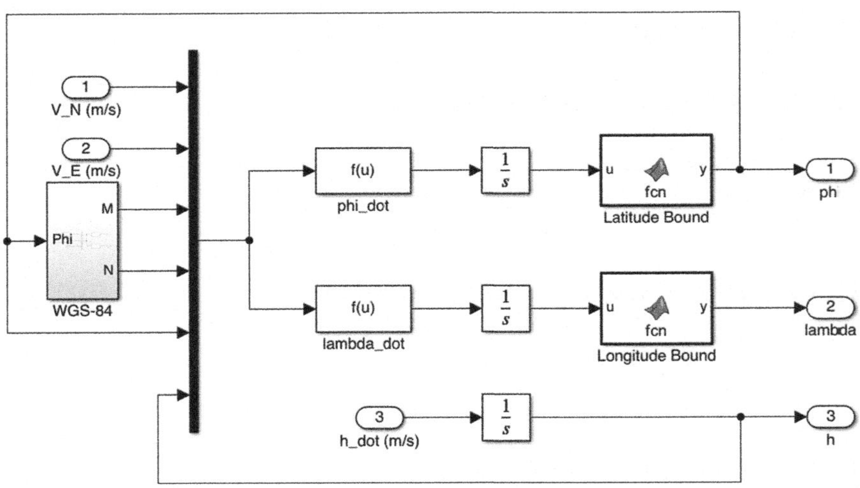

Figure 13.3 Geodetic position of the aircraft in Simulink.

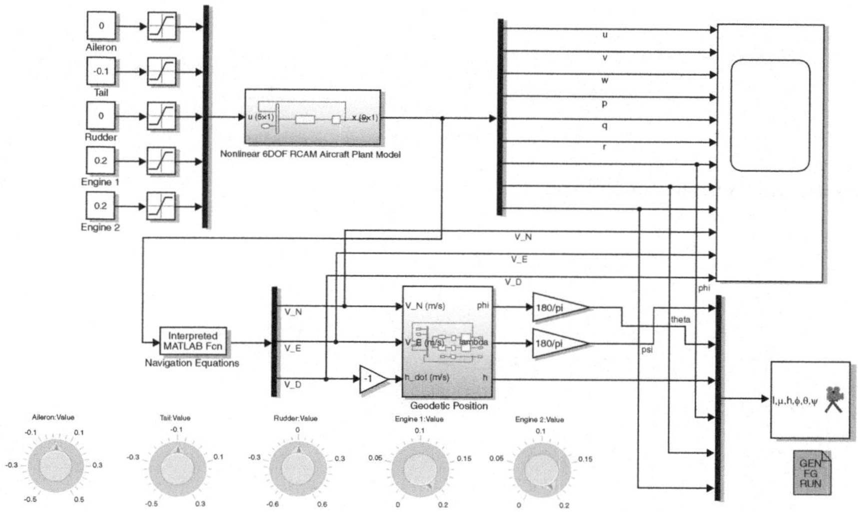

Figure 13.4 Complete block diagram in Simulink of the Research Civil Aircraft Model (RCAM).

13.11 COMPLETE BLOCK DIAGRAM IN SIMULINK

The figure above shows what the complete block diagram looks like within Simulink. The controls for the aircraft can be set as constants or adjusted via the knobs at the bottom. Outputs can be accessed using the scope to view the state values in a graph, or by using FlightGear. It is important to note that the values used in the model are all in radians, while the simulator only accepts degrees; this is why a multiplier of $180/\pi$ is applied before the inputs enter the simulation.

13.12 RESULT AND DISCUSSION

Figure 13.5 shows the initial conditions for each control. The corresponding values for the aileron, tail, rudder, engine 1, and engine 2 are 0, –0.1, 0, 0.2, and 0.2, respectively. These values serve as the reference conditions for the simulation.

Figure 13.6 illustrates the waveform of the state variables of the aircraft modeled in Simulink. These output values demonstrate the aircraft's responses to the initial conditions specified in Figure 13.6. Each graph's analysis is detailed based on its respective column group.

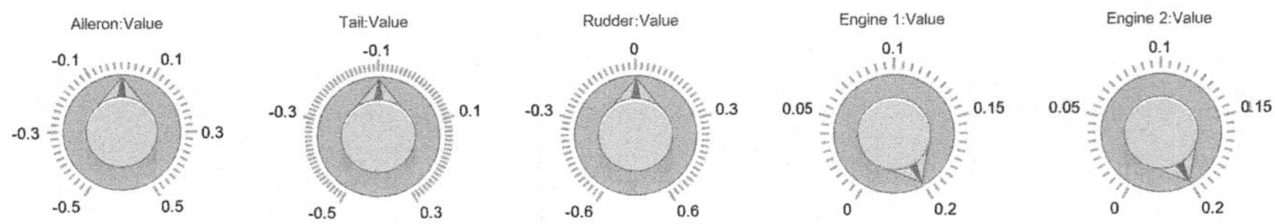

Figure 13.5 Aircraft control knobs.

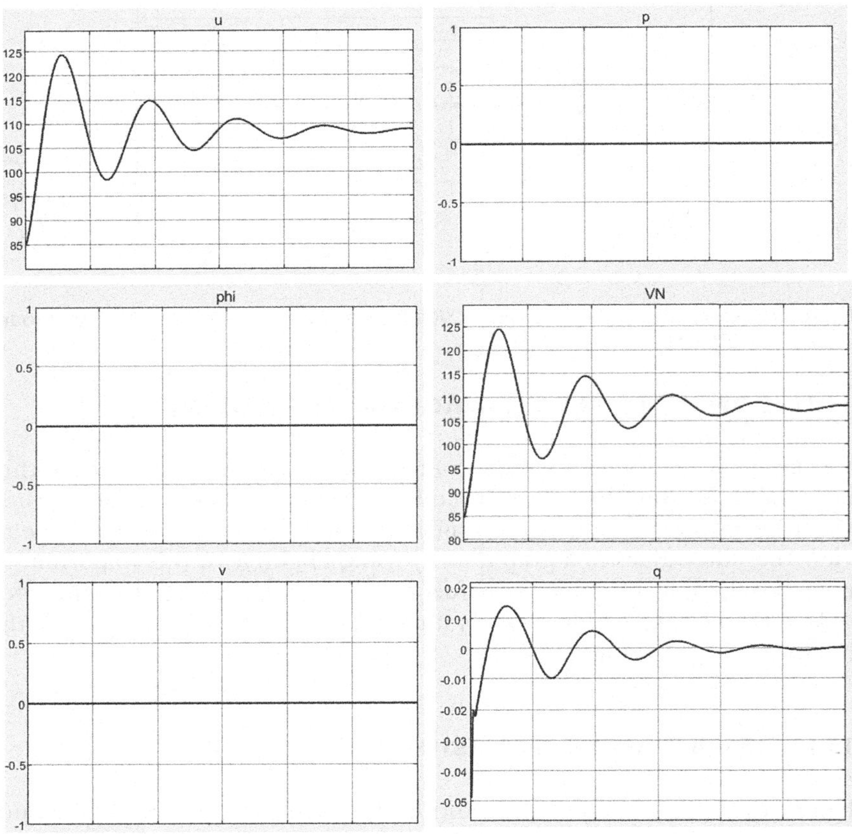

Figure 13.6 Scope of state variables.

(Continued)

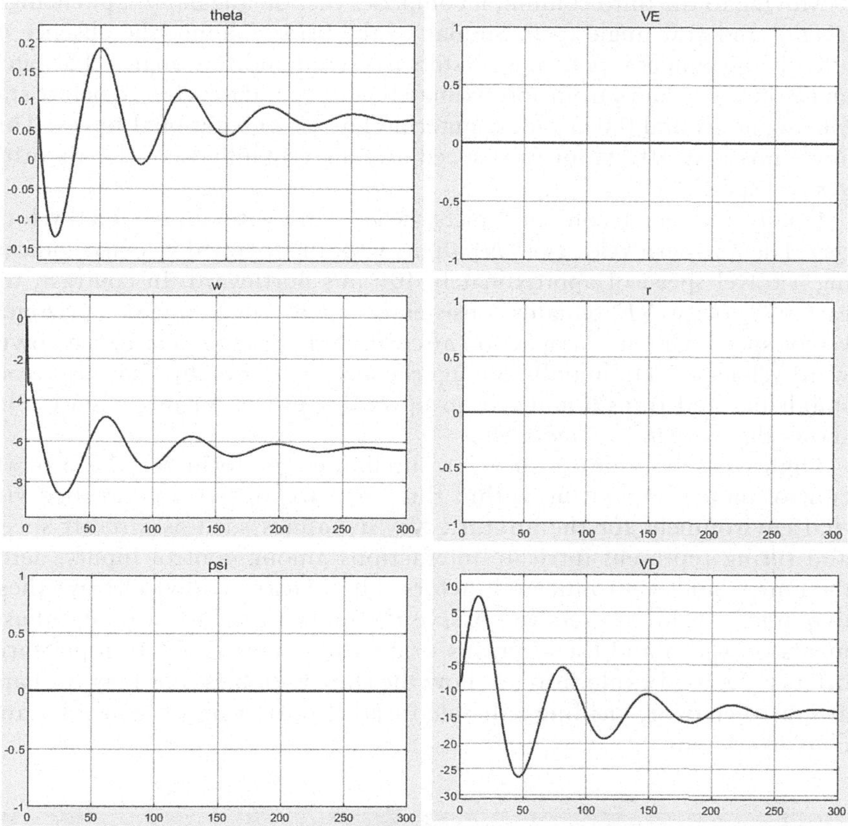

Figure 13.6 (Continued) Scope of state variables.

In the first column, the longitudinal velocity (*u*) initiates at 85 m/s, reaching a peak of approximately 124 m/s. Oscillations are evident in its graph as the forward speed fluctuates due to pitch and gravity. It eventually stabilizes at around 109 m/s, a result of the aircraft's inherent stabilization characteristics. Meanwhile, the lateral velocity (*v*) remains consistently zero as there are no changes in the aileron and rudder controls. Conversely, the vertical velocity (*w*) starts at 0 m/s and oscillates before settling at a steady –6 m/s.

The second column portrays the aircraft's angular velocities, consisting of the roll rate (*p*), pitch rate (*q*), and yaw rate (*r*). With no sideward force, the roll and yaw rates of the aircraft remain at zero. However, a pitch velocity is evident as the aircraft endeavors to stabilize. Beginning at –0.05 m/s, it oscillates before eventually reaching a state of rest at zero.

Moving to the third column, it comprises the roll angle (*phi*), pitch angle (*theta*), and yaw angle (*psi*). Similar to the prior column, the absence of aileron and rudder input values results in no roll and yaw angles. The pitch angle theta oscillates from approximately −0.125 rad to 0.18 rad before stabilizing at around 0.075 rad, countering Earth's gravitational forces. This angle inversely affects forward speed and directly influences the aircraft's vertical speed.

Finally, the fourth column depicts the aircraft's actual directional velocities. The northern velocity (*VN*) aligns with the forward velocity, indicating a travel speed of approximately 109 m/s northward. In contrast, the eastern velocity (*VE*) remains consistently at zero, suggesting no horizontal vector speed. Gravity's impact on aircraft models is apparent in the downward velocity (*VD*), initially pulling the aircraft at roughly 7.5 m/s before stabilizing and transitioning to an upward speed of 14 m/s. Figure 13.7 shows the *Aircraft in FlightGear.*

FlightGear provides a superior simulation platform for the aircraft control model. Operating within this software creates a controlled virtual environment for the aircraft. Key dynamics such as aircraft speed and tilting represent intricate interactions among control inputs, aerodynamics, and environmental factors. FlightGear vividly portrays these dynamics, enabling users to observe the impact of control input adjustments on speed and the aircraft's orientation. This simulation platform offers a detailed exploration of how the state variables interplay to shape aircraft behavior, yielding valuable insights into aviation principles and control systems.

Figure 13.7 Aircraft simulation in FlightGear.

13.13 APPLICATIONS

This section explores the applications of an aircraft model, specifically the RCAM created using MATLAB and Simulink.

13.13.1 Interactive learning

It allows researchers and students to analyze how the accuracy of mathematical representations influences the actual behavior of the aircraft in the real world. In addition, this can also be utilized to understand the impact of control algorithms on the aircraft's stability during maneuvering.

13.13.2 Simulation-based training

The RCAM offers an aircraft flying experience in a realistic environment. It is a valuable tool that provides a practical, hands-on approach to understanding aircraft operations and can serve as training before operating a real aircraft.

13.13.3 Aerospace research and development

The RCAM is a specialized aircraft model, similar to the Boeing 757-200, designed to simulate a twin-engine civil aircraft. This can serve as a platform for testing new concepts and designs on aircraft before implementing them in real-world scenarios.

13.14 CONCLUSION

This laboratory experiment utilizing MATLAB, Simulink, and FlightGear for the RCAM simulation provides comprehensive insight into aircraft dynamics and control. The simulation revealed responsive interactions among control inputs, aerodynamics, and environmental factors, notably observed in the aircraft's state variables and responses. Analyses of the state variables showcased how different inputs and conditions affected the aircraft's behavior. FlightGear, in particular, proved instrumental in vividly representing these dynamics, offering a realistic platform to observe the impact of control adjustments on speed and orientation [8–12]. This study not only demonstrates the versatility of simulation tools in understanding complex aircraft systems but also highlights their potential in studying aviation principles and refining control mechanisms for enhanced aircraft performance and safety.

REFERENCES

1. Bryan, V., Petrauskaite, G., Charpentreau, C., & Stephenson, R. (2022, July 29). 10.real-world examples of why aviation is essential. *AeroTime*. https://www.aerotime.aero/articles/31760-ten-examples-when-aviation-becomes-essential
2. Lum, C. (2020). A nonlinear, 6 DOF dynamic model of an aircraft: the research civil aircraft model (RCAM). Retrieved December 2, 2023, from https://youtu.be/bFFAL9lI2IQ?si=xBVrWuqTRmsCmT6m
3. Prameswari, A. W., Satyawan, A. S., & Ariffin, D. M. (2021). Research civil aircraft model (RCAM) dengan metode dinamis 6 derajat Kebebasan Menggunakan matlab. *Prosiding Seminar Nasional Sains Teknologi dan Inovasi Indonesia (SENASTINDO)*. https://aau.e-journal.id/senastindo/article/view/152
4. FlightGear Flight Simulator. (n.d.). Introduction to flightgear. *FlightGear Flight Simulator*. https://www.flightgear.org/about/
5. MathWorks. (n.d.). Simulink – simulation and model-based design. *Simulation and Model-Based Design - MATLAB*. https://www.mathworks.com/products/simulink.html
6. Arora, K., Kumar A., Kamboj, V. K., Prashar, D., Shrestha, B., & Joshi, G. P. (2021). "Impact of renewable energy sources into multi area multi-source load frequency control of interrelated power system". *MDPI Journal of Mathematics*, 9(2), 186 (2021). https://doi.org/10.3390/math9020186/
7. Konz, C., Happel, C. C., Turano, D., Daniel, G., Bigger, M., Ratliff, O., & Leishman, J. G. (2023, January 1). Aircraft classifications & regulations. *Introduction to Aerospace Flight Vehicles*. https://eaglepubs.erau.edu/introductiontoaerospaceflightvehicles/chapter/aircraft-classifications-aviation-regulations/
8. Pal, V. C., Tripathi, S. L., & Ganguli S. (March 2024). *Industrial Control System, Wiley*. https://doi.org/ 10.1002/9781119829430, ISBN: 978-1-119-82925-6
9. Tripathi, S. L. (2024) Dynamic modeling of drone control with MATLAB Simulation05 September 2024. https://doi.org/10.1002/9781119847533.ch5
10. Roy, T., Tripathi, S. L. &Ganguli S. (2024). *Linear and Nonlinear System Modeling*, Wiley Scrivener Publishing LLC. https://doi.org/10.1002/9781119847533
11. Sahani, R., aga, M., Jamatia, I., Kumar, P. & Tripathi, S. L. "Design of a self-charging electric vehicle with embedded multiple natural energy sources," *2023 IEEE Devices for Integrated Circuit (DevIC)*, Kalyani, India, 2023, pp. 296–301. https://doi.org/10.1109/DevIC57758.2023.10134774
12. Kumari, C.L., Kamboj, V.K., Bath, S.K. et al. (2023). A boosted chimp optimizer for numerical and engineering design optimization challenges. *Engineering with Computers* 39, 2463–2514. https://doi.org/10.1007/s00366-021-01591-5

Index

For Product Safety Concerns and Information please contact our EU representative GPSR@taylorandfrancis.com Taylor & Francis Verlag GmbH, Kaufingerstraße 24, 80331 München, Germany

Printed and bound by CPI Group (UK) Ltd, Croydon, CR0 4YY
14/09/2026
02216256-0010